William Henry Ford

Boiler Making for Boiler Makers

A practical treatise on work in the shop

William Henry Ford

Boiler Making for Boiler Makers
A practical treatise on work in the shop

ISBN/EAN: 9783744758451

Printed in Europe, USA, Canada, Australia, Japan

Cover: Foto ©berggeist007 / pixelio.de

More available books at **www.hansebooks.com**

BOILER MAKING

FOR

BOILER MAKERS.

A PRACTICAL TREATISE ON WORK IN
THE SHOP.

SHOWING THE BEST METHODS OF
RIVETING, BRACING, AND STAYING, PUNCHING,
DRILLING, SMITHING, ETC.

AND

THE MOST ECONOMICAL MANNER OF OBTAINING
THE BEST QUALITY OF OUTPUT AT
THE LEAST EXPENSE.

BY

W. H. FORD, M.E.

NEW YORK:
JOHN WILEY & SONS,
15 ASTOR PLACE.
1887.

CONTENTS.

CHAPTER		PAGE
I.	MATERIALS	9
II.	TESTING MATERIALS	14
III.	BOILER FORMS	20
IV.	RIVETED JOINTS	44
V.	BRACING AND STAYING	58
VI.	FLANGING	92
VII.	WELDING PLATE	108
VIII.	ANNEALING	115
IX.	SMITHING	121
X.	PUNCHING	138
XI.	DRILLING	150
XII.	TRIMMING	155
XIII.	COLD BENDING	159
XIV.	SETTING UP	164
XV.	CALKING	175
XVI.	TUBE SETTING	181
XVII.	FITTINGS	189
XVIII.	TESTING	196
XIX.	ORDERING STOCK	204

PREFACE.

Shortly after entering the profession of mechanical drafting, I was given a boiler to design. That is to make a working drawing. I had been taught the principles as to the capacity or size in relation to the power, the thickness of shell required to resist the strains, and other such generalities which relate to types and relative economy, but barely a hint was given me as to the multifarious small details which go to make up the difference between work made for sale and that made for service.

In consequence of my lack of knowledge, I was compelled to consult the foreman boiler-maker oftener than I should. He being as kind as he was experienced, generously gave me the required information for that boiler. This of course not being sufficient for my wants, I set about looking for works on the subject. My disappointment was great, indeed, to find that among the very few books on the subject, there were absolutely none which gave the information that I desired. Such as were complete buried themselves over neck and ears in abstruse formulas which it is doubtful if the authors could solve after a month's absence from the work.

I commenced making notes of any and every thing on the subject; not only the details, but the methods of making them. Having shown my notes to a foreman boiler-maker he advised me to write them out, and have them printed, for it was just such matter as every boiler-maker wanted. This having been done and shown to several boiler-makers, further encouragement was given to go ahead.

If any should ask, why the power or capacity and economy of boilers is not touched upon, I would say that that is strictly in the field of Engineering, which apparently is well filled, while this book is entirely devoted to the Workshop. At the same time it is to be hoped that some engineers, at least those just entering into the workaday world, may find something that will help them while relieving the boiler-makers of many disagreeable and difficult constructions. Again it may be asked why then I introduce strength of materials and such matters. In answer I would say that nine out of ten drawings sent to the shop, leave such things entirely to the boiler-makers, who, more than half of them, have no time to study Rankine or mathematical science, even were they so disposed.

The endeavor, in this work, is to make it one of ready reference, at the same time enabling one to work out for himself, in a simple way such problems as would naturally come up in the course of his daily work.

W. H. FORD.

December 23, 1885.

I.

MATERIALS.

Practically, wrought iron and steel are the only metals used in boilers. Some few have brass or copper tubes where wood is used for fuel. Copper fireboxes have long been discarded in this country. The extra thickness required for strength as well as to allow for the rapid deterioration caused by the action of coal gasses more than making up for its greater power of conduction.

In boiler plates the quality of tensile strength should not be the sole consideration. The mere fact that a piece of unworked plate, pulled apart in a testing machine, resisted so many thousand pounds, is no proof that it will withstand the various processes of flanging, punching, bending, etc., in the shop, nor the constant and varying strains, from heat and cold, in service. Ductility and toughness are as essential qualities as great tensile strength.

"Steel" plates should have a tensile strength of 55,000 to 60,000 pounds per square inch of section. If the strength goes much beyond this the excess of carbon required to produce it causes brittleness and makes it difficult to work. In a test piece, four inches long, the elongation should be not less than 25 per cent. Boiler steel goes under various names or brands, each maker having his own, and claiming special points of excellence for each. However, nearly all will meet the above named requirements.

Iron plates have several brands, not peculiar to any one maker, but used by all, that are intended to indicate the quality. These are:

    C. H. No. 1, Firebox,
    C. H. No. 1, Flange,

MATERIALS.

Shell,
Refined, and
Tank,

C. H. No. 1 Firebox and C. H. No 1 Flange are made entirely of charcoal iron. The C. H. No. 1 Firebox is somewhat harder than the Flange, so as to resist more effectually the fierce heat of the fire. They should stand from 50,000 to 55,000 pounds tensile strain in any direction if the makers use the brand honestly.

"Shell" iron has but very little charcoal iron in its composition and that as a skin. It cannot be flanged. Neither can it be bent across the grain. It should stand, with the grain, 45,000 to 50,000 pounds; while across the grain 35,000 is good.

"Refined' is rather a doubtful term, and really means that the iron is refined from the pig It may be that it is simply muck bar or that it is a fair quality, which has undergone several workings.

The actual quality is best determined

by the price. The more the iron is worked the better it is and of course the greater the price.

"Tank" iron is a poor quality which will hold water, but not under much pressure. It cannot be flanged. With the grain in light plates it can be bent to about a four inch radius.

There are special brands of extra good qualities of iron such as Sligo, N. P. U., Eureka and Pine. These are almost the same as the so-called homogeneous steel in their composition. The tensile strength is the same, but the percentage of elongation is less, about 15 to 20 per cent.

If there are no means of determining the strength of plates in boilers, the U. S. Government inspectors rate them at 45,000 pounds to the square inch. The maker should always brand the plates with his name and the amount of tensile strain he thinks it will safely bear. This brand should always be placed in the boiler in such a position that it can readily be found. Boiler-makers not having a testing machine of their own, should require a guarantee of the plate makers. This should be kept, and an accurate record made of the disposition of the plate described by it.

Iron for stay bolts should be of the very best, strong, tough and ductile. It should show a tensile strength of not less than 48,000 pounds per square inch, and have an elongation of at least 28 per cent. in a 4-inch specimen. It should cut a clean thread with the dies in fair condition.

For rivets, it is important that the metal should possess the qualities of toughness and ductility in the highest degree. An immense strain is thrown upon them by their contraction in cooling, when sooner or later, if the metal is short or brittle, they will snap with more or less dangerous result. At present rivets are mostly made of iron, but steel seems to be the coming metal. Already the British Admiralty have demanded it, and are using it with good results. It is required to be of a tensile strength of 58,000 to 67,000 pounds per square inch and have an elongation of not less than 20 per cent. in a length of 8 inches.

Iron for stay rods should be of good quality, having a tensile strength of not less than 40,000 pounds per square inch.

Cast iron is used in boiler work to a very limited extent. It makes very good thimbles and feet for crown bars, frames and covers for man and hand holes. It is sometimes used for the heads of plain cylindrical boilers. But it is rather unreliable and should never be used where its tensile strength is to be depended upon.

II.

TESTING MATERIALS.

As soon as possible after receiving a lot of material it should be closely examined and tested for size and soundness. There are so many chances of error in size, and of defects that may condemn, that to avoid delay and expense the sooner such are discovered the better. All material should be tested for tensile strength.

The thickness of plates should be particularly looked after. If it is over 5 per cent. less than the thickness called for it should be rejected, as it will materially affect the strength, as well as be an aggravation to fit. An excess of 5 per cent. should also cause rejection, or at least a rebate on cost to the amount due for over weight. Examine the surface and edges thoroughly for laminations, blisters and other flaws, such as cracks, indentations or marks caused by careless rolling. To determine its internal soundness, set it up on edge and go over it carefully with a light hammer. If the blow gives a sharp, ringing sound, the plate is good, but if it is dull and heavy in character, it indicates a defect. To be sure of trying all parts of the surface the plate may be lined off into squares of four to five inches. To still further test the soundness, it may be slung by the corners horizontally, and strewn with dry sand. On being lightly tapped underneath, the sand will be thrown off where the plate is solid, but will remain fixed where a blister or lamination occurs. While this is being done, a strip from the edge should be tested for tensile strength.

If the plate passes the foregoing, it should then be subjected to the working tests. A strip should be taken from the plate, $1\frac{1}{2}$ to 2 inches wide, and bent cold over the corner of an anvil. This corner should not be of more than $\frac{1}{2}$ inch radius. A $\frac{1}{4}$ inch plate of good iron should bend, with the grain, to an angle of 90 degrees without cracking, and across the grain to an angle of 55 degrees. The angle for heavier plates will be as per table:

Thickness of plates..	$\frac{1}{4}$	5-16	$\frac{3}{8}$	7-16	$\frac{1}{2}$	$\frac{5}{8}$	$\frac{3}{4}$	$\frac{7}{8}$	1.
To bend across the grain to (degrees)..	55	52	49	46	43	37	31	25	20.
To bend with the grain to (degrees)	.90	85	80	75	70	60	50	40	30.

Flange iron, when heated to a bright red, should bend double without cracking at the edges.

Steel should bend double when cold and at a cherry red heat. Also after being heated to a cherry red and cooled in water at a temperature of 80 degrees.

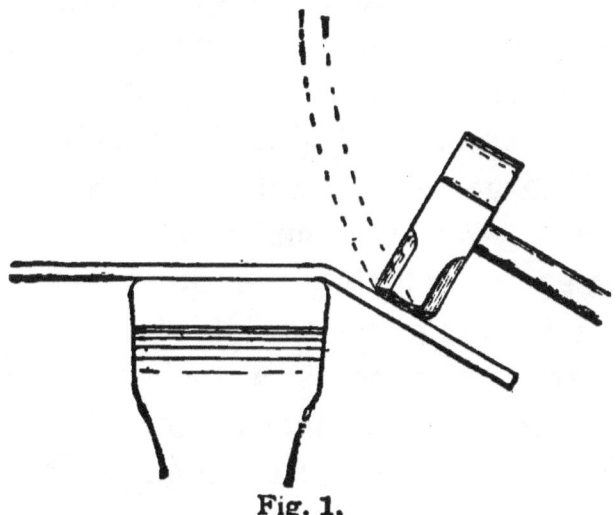

Fig. 1.

The parts hammered down so that daylight cannot be seen between them. Be careful in heating steel to bring up the heat slowly. A quick fire is apt to burn the surface before the centre is fully heated. If the metal will bear these tests without cracking at the edges, it shows a very good quality that will work well and easily. Bending can be done more successfully by striking a glancing blow. See figure 1.

Tubes should be examined for seams, flaws in the welding and flat spots. As they are subjected to external pressure this last is a fatal defect. Being thin they are very easily flattened and should be carefully handled. A tube should stand expanding tightly in a hole 1-16 inch larger than itself, and the edge turned over without cracking transversely or splitting. A good test is to take a short piece say 1¼ inches long, where the tube is 2 inches outside diameter, and hammer it down flat as in figure 2. When down solid it should

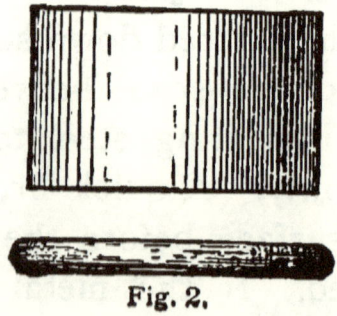

Fig. 2.

show no transverse cracks and but slight splitting.

Staybolt iron, beside testing for tensile strength, should also be tested by bending double till the two parts meet. It should then show no cracks.

Rivets are seldom tested for tensile strength. About one in a hundred are usually taken for forge testing. They should stand bending cold to the shape shown in figure 3. The space *a* to be

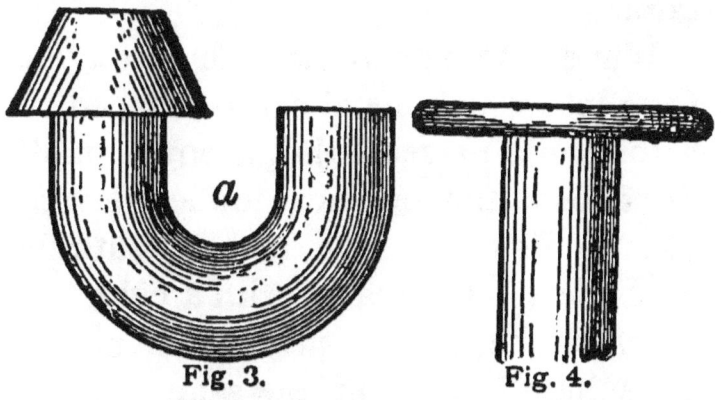

Fig. 3.   Fig. 4.

not over one diameter of the rivet. Bent hot they should close the parts together. In neither case should they show a crack. No cracks should appear on the edge of the head when flattened out, hot, to $\frac{1}{8}$ inch thick, as in figure 4.

III.

BOILER FORMS.

The forms of boilers and their parts are various, though composed mostly of cylinders or combinations of other forms with cylinders. The cylinder being the only form adapted to the purpose that is inherently self-sustaining.

While the sphere is the strongest form into which metal can be shaped to withstand a pressure that is equal in all directions, the practical difficulties of construction preclude its use entirely for boilers. Compared with a cylinder, in which the length equals the diameter, the sphere has about the same amount of available heating surface ; while the cubic contents is one-third less.

Next to the sphere in strength comes the cylinder with hemispherical ends, or egg-ends as they are called in England. These ends are costly and difficult to make, and as the flat ends are very readily stayed, they are far more preferable.

To find the thickness of a hollow cylinder to withstand an internal pressure take:

$$t = \frac{D \times P}{2c} \times K \dots\dots\dots\dots 1$$

in which $t=$thickness in inches,
$D=$diameter " "
$P=$pressure per square inch in pounds,
$c=$tensile strength, per square inch of section,
$K=$factor of safety.

This factor of safety is an arbitrary figure, for which in boiler-making 6 is usually taken. This nominally makes the boiler six times as strong as it really needs to be, but as it is used in this case, it only applies to the strength of the plate and is intended to cover the weakening effect of bad workmanship, flaws and riveted joints. The strength of these joints ranges from 80 per cent. of the solid plate down to as low as 50 per cent.

The strength of a cylinder to resisbursting may be found by taking :

$$S = \frac{2\,t \times c}{D \times K} \dots\dots\dots\dots\dots 2$$

The thickness t is taken twice because there are two sides of the cylinder to resist the pressure. Take any opposite points as $a\ b$ in figure 5. Now the total pressure to be resisted at these two points, for a section 1 inch long, is the pressure per square inch multiplied by the distance they are apart in a straight line, (or the diameter).

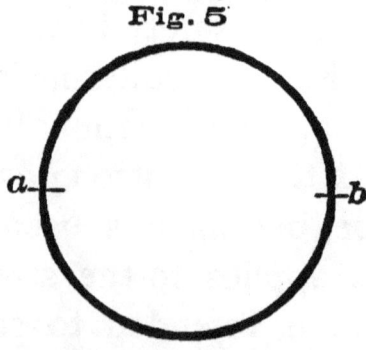

Fig. 5

To prove that this strain is due to the diameter and not to the circumference, take *a c b* in figure 6, as one-half

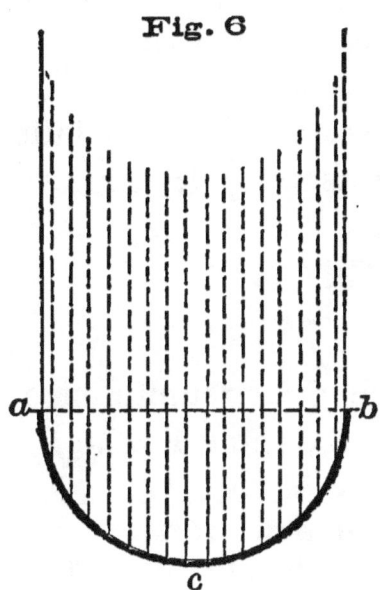

Fig. 6

of the cylinder. Divide the diameter into a number of equal parts. Draw parallel lines at right angles to *a b*, from *a c b*. These lines show the direction of the force tending to rupture the plate at the points *a* and *b*. Now extend these lines beyond *a b*, making them all of a length, by scale, equal to the pressure in lbs. per square inch.

Then if the sum of these lines in pounds per square inch be multiplied by the distance in inches between them, the answer will be, as before stated, the total amount to be sustained by the two sides.

For the strength of a cylinder to resist the force tending to rupture it transversely take:

$$S = \frac{C \times t \times c}{F} \quad \ldots \ldots \ldots \ldots 3$$

in which C=circumference in inches. And for the force F take:

$$F = A \times P \quad \ldots \ldots \ldots \ldots 4$$

in which F=force
A=area of end.

A short calculation will show at once, that if the cylinder is strong enough to resist a diametrical force, that it will be several times strong enough to resist a longitudinal one.

It must be remembered that flat ends require thorough bracing, which will be considered in full under the head of "Bracing and Staying."

While an internal pressure tends to produce and retain a circular, or more properly, a spherical shape, an external pressure has a directly opposite tendency. Therefore cylindrical flues and tubes to be self-sustaining under external pressure must be truly circular in section. If not so, the only dependence for resistance to the pressure is the stiffness of the plate. For example take a 30 inch diameter flue, which has been flattened so that it forms an oval, like figure 7.

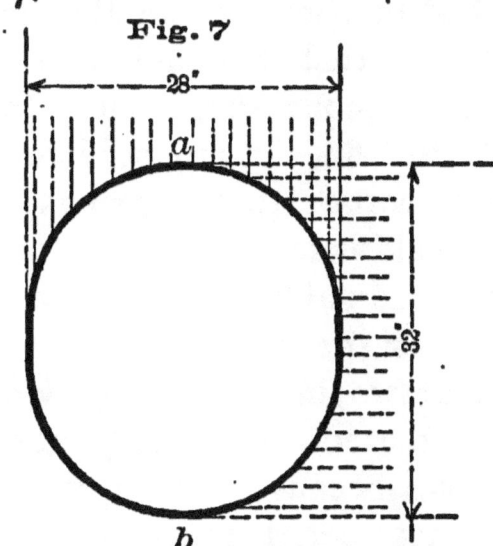

It will be readily seen that the total

pressure upon the flue on the long side is thirty-two times the pressure per square inch; while upon the short side the total pressure is twenty-eight times the pressure per square inch. Now if the pressure was 100 pounds per square inch, then the preponderance of pressure = (32—28) × 100 or 400 pounds. To resist this 400 pounds, we have only the stiffness of the plate from *a* to *b*. If this 400 pounds should spring the plate in the least, it would add to itself and take away from the short side as much, and this in a constantly increasing ratio, so that it would take but a short time to collapse the flue. In this example the flue is supposed to have no end support or stiffening rings, which materially aid it to resist distortion.

A longitudinal lap joint should be avoided, as the lap cannot be made without destroying the symmetry of the circle. A butted and welted joint is the proper one, because with it, the cylinder is as near perfect in its shape as it is possible to make it, and is almost entirely independent of the welt for help in preserving its form. About the only duty of the welt is that of a "stop-water," and for this a single one placed outside is sufficient.

For the thickness of plate in cylindrical flues, where the length does not exceed fifteen times the diameter, take:—

$$t = \sqrt{\frac{L \times D \times P}{806,300}} \dots \dots \dots \dots 5$$

in which L=length in feet.

More properly and in accordance with an exhaustive series of experiments by Sir. Wm. Fairbairn, it should read:—

$$t = \sqrt[2.19]{\frac{L \times D \times P}{806,300}} \dots \dots \dots \dots 6$$

But, as a factor of safety of five or six is or should be allowed, the former is sufficiently accurate and is far simpler.

The transverse joints of flues should be made by flanging out the ends of the sections and riveting between them a solid forged ring as in figure 8.

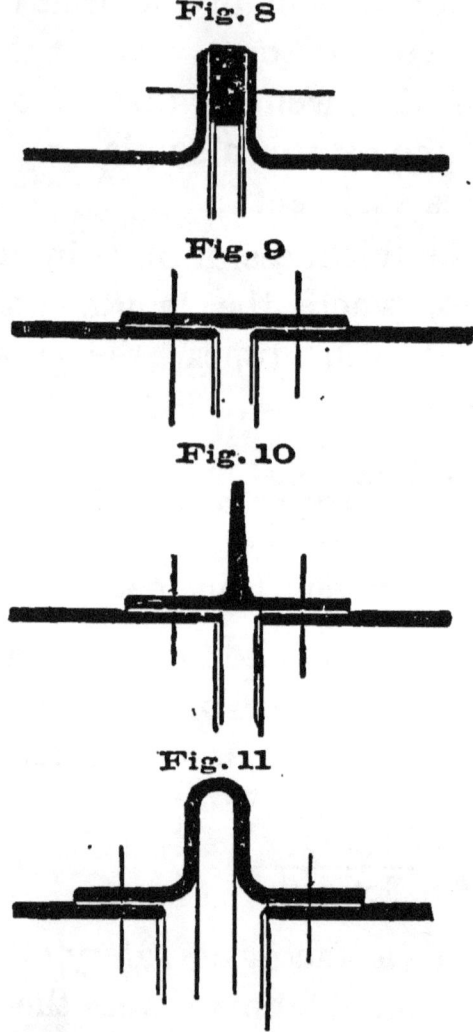

Fig. 8

Fig. 9

Fig. 10

Fig. 11

This ring should not be less than ¾

inch thick, to allow for calking. Nor should the inside radius of the flange be less than twice the thickness of the plate, to allow for the difference of expansion of the flue and shell.

Other methods of forming girth joints of flues are as shown in figures 9, 10, and 11.

These are all very expensive to fit as they necessitate the greatest care in having the ends of the sections exactly the same diameter, and to have the rings *drive* on. If this is not done the joint will be an infinite source of trouble from the very first. Figure 11, although the most expensive of all, is an ideal joint, and is always used in extra high class work. Figures 9 and 10 do not allow for expansion, and should not be used unless other joints like 8 or 11 are used at least once in the same flue. Figure 10, although often used, is very bad. Rolled tee-iron when bent, either hot or cold, is very apt to split at the root of the flange, and should be condemned altogether for any purpose whatever in a boiler. In any of these joints, the ends of the plates should have not less than an inch between them to allow for calking.

Small lap-welded tubes are made much thicker than is required for strength to resist the pressure put upon them. The rapid wasting from corrosion, and the wear from the sand blast action of the cinders, necessitating a much greater thickness. The difficulty of making a successful weld also keeps up the thickness. A very thin plate loses its heat too quickly.

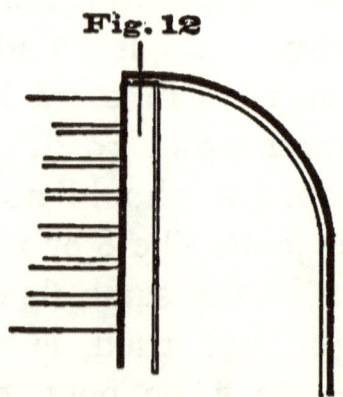

Fig. 12

No dependence should be put upon

forms that are segments of cylinders, to sustain themselves. Such forms require staying as well as flats, although not in so great a degree. See figure 12, which is a common arrangement of the back connection in marine boilers.

Where large openings are made in the shell of a cylindrical boiler, such as for manholes, domes, etc., compensating rings should be used to make up for the loss of section. It must be constantly kept in view that such openings take away that much of the cylindrical shape as well as throwing so much more labor upon the adjoining portion of the plate.

A good method of stiffening under a dome is by butting the barrel sheet on

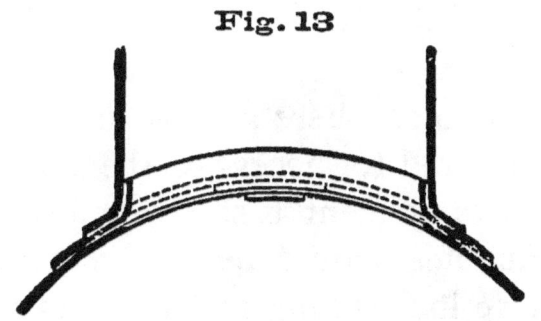

Fig. 13

the same longitudinal centre. The outside welt being made wide enough to follow around the opening with a double lap and flanged up into the dome, as in figure 13.

This welt should be 25 to 50 per cent heavier than the shell. The inner welt, which need not surround the hole, may be made in the usual manner. Figure 14 shows a method largely used in locomotive boilers, which is not only a difficult job to fit and rivet up solidly, but is very unreliable.

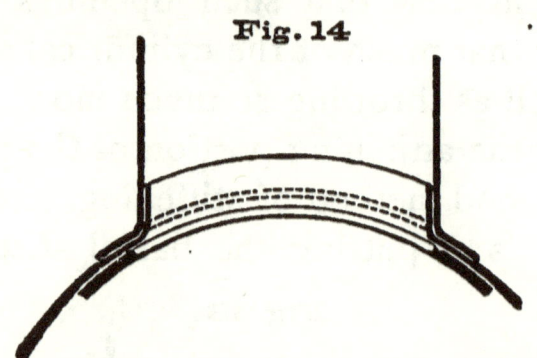

Fig. 14

For domes where a large opening is not required for access to the boilers a good arrangement is shown in figure 15.

Manholes should be oval and about 15 or 16 inches long and 12 or 13 wide.

The long axis to be placed transversly to the boiler. The frame should be

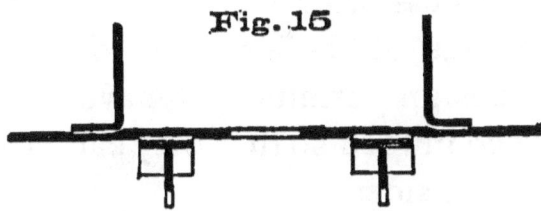

Fig. 15

amply stiff to compensate for the part of the shell cut out, and should be inside the boiler. The plate also should be inside and have two handles. The joint between plate and ring is better if faced, but this is not absolutely necessary when a rubber gasket is used, if it is fairly straight.

Flat surfaces are not sufficiently self-supporting, unless made enormously thick. To find the thickness of a circular head take:

$$t = \sqrt{\frac{D^2 \times P}{4c}} \quad \ldots\ldots\ldots\ldots 7$$

This shows what a heavy plate is required without allowance of a factor of safety. The best way is to allow it the same thickness as is in the rest of the boiler and stay it, as if it had no strength or stiffness whatever.

Heads of boilers should, wherever the design permits of it, have the flange turned inward so that they can be calked on both sides.

Fig. 16

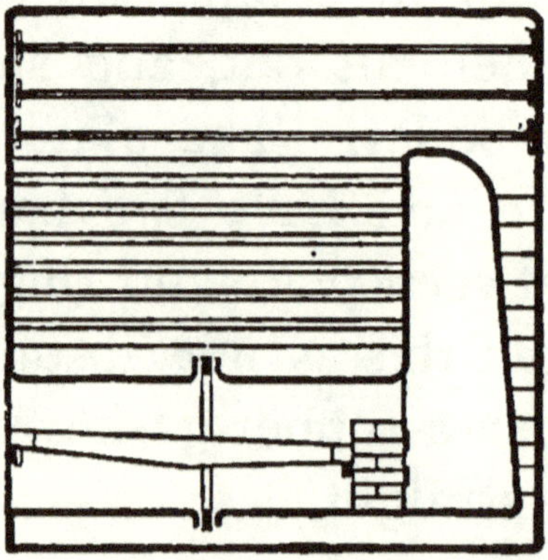

Rectangular boilers with their great expanses of flat surfaces are fast becoming obsolete. The pressures used of late years being too great for them without overloading them with braces and bars, till they are almost inaccessible for cleaning out and repairs. The circulation also is badly obstructed.

Furnaces are usually of the same

general form as the shells. For cylindrical boilers they are cylindrical. Their strength to resist pressure, and the details of their joints, are the same as described for flues. In marine return tubular boilers they are connected with the tubes above them by a vertical flue called the back connection. See figure 16.

The longitudinal joints of these furnace flues should be entirely below the grates so as to keep them out of the intense heat of the fire.

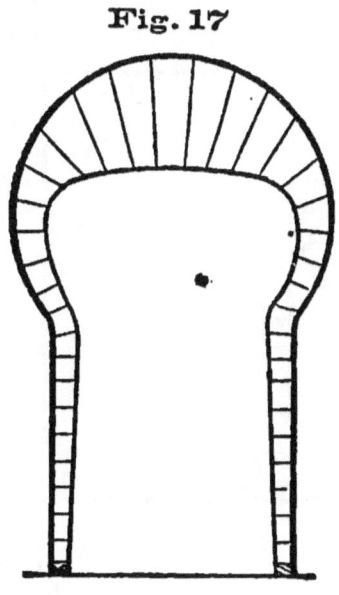

Fig. 17

BOILER MAKING.

Locomotive furnaces or fire-boxes, as they are commonly called, are mostly made with a flat or slightly arched crown, although quite a percentage have a semi-elliptic form, the latter being used with a straight top boiler, and the former with the raised wagon top. This raised wagon top gives considerable flat surface to stay, at the point *a* figure 18. Sometimes this flat sur-

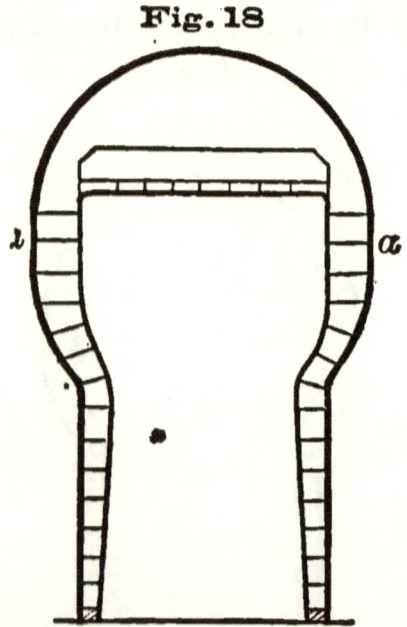

Fig. 18

face is gotten rid of by making the side of a large radius, say from one and one-half to twice the radius of the top, producing an óval, in the fond hope of making it partially self-supporting. It only makes a neater appearance without enhancing its strength or stiffness to any practical amount. Nor as long as it is not a true circle, and a complete one at that, can it be done. The semi-elliptic crown is stayed with through bolts, the flat or slightly arched crown with girders, except in the Belpair type (fig. 19)

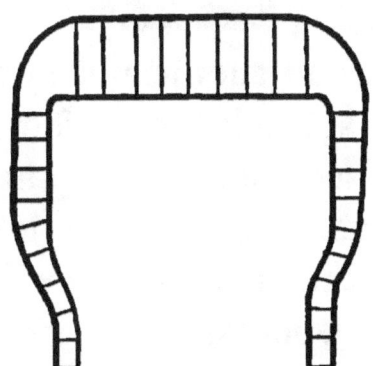

Fig. 19

where the shell is brought down flat, in which through bolts are used. It is desirable in narrow water spaces, as at the sides of locomotive fire-boxes, and the back connections of marine boilers that the space should be widened at the top as much as possible, so that the ascending steam which accumulates as it rises may have a greater freedom to escape into the steam-room without blowing the water out of the leg.

Where the space is large enough, the bottoms of water legs are formed by flanging up a plate as in figure 20.

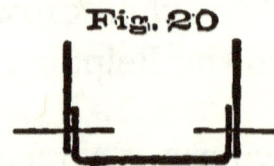
Fig. 20

This method should not be attempted where the space is not sufficient to get a good sized holding-on hammer or block against the rivet. It is especially difficult in the narrow space allowed in locomotive boilers. For them the arrangement shown in figure 21 is used.

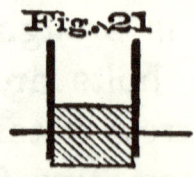
Fig. 21

The bar should not be less than $2\frac{1}{4}$ inches thick, and because of the difficulty in thoroughly upsetting the rivet at both ends; it should not be over 3 inches wide.

A width of two inches would be far better, and to provide for good circulation the space can be rapidly widened above. Another plan is to flange the inside sheet in the form of an ogee to meet the outside one as in figure 22.

Fig. 22

But this beside being difficult to calk well has the great disadvantage of grooving caused by deposits of sediment, which it is almost impossible to remove from the sharp angle between the plates. This deposit will rapidly harden and form a fulcrum for the plate to work upon, which is rapidy taken advantage of, and the plates become dangerously weak. Care must be taken to set the grates well above these joints, as the intense heat of the fire will soon burn the heads off the rivets as well as the sheet if there is much accumulation of sediment.

For furnace-doors, figures 26 and 21 are used as well as figures 23, 24, and 25; which are much better. Figure 25 has proven to be the best of all, taking all things into consideration.

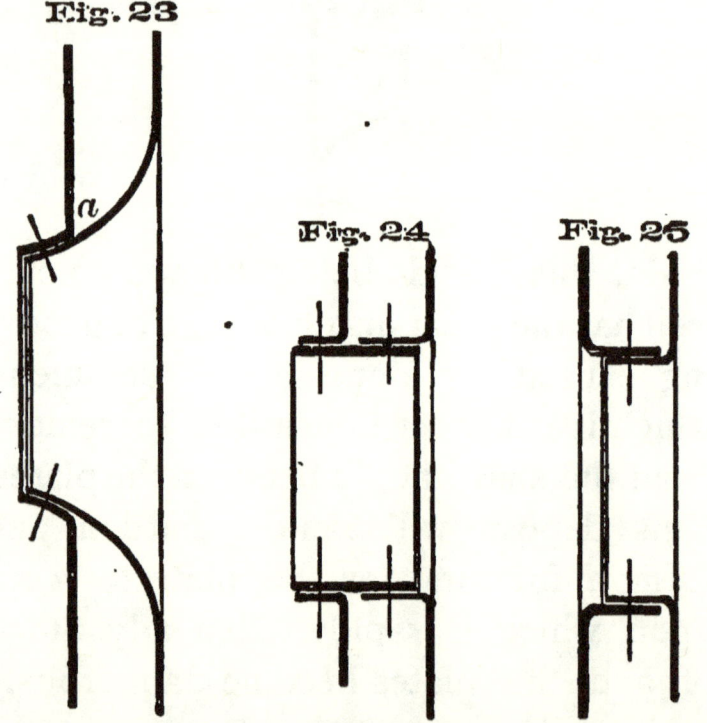

Figure 23 is the most convenient for working the fire, but gives more or less trouble from grooving at *a*. Figure 24 is entirely too deep for firing properly at the back end' and is costly and difficult to fit.

Combustion chambers are often used in locomotive and other internally fired boilers, to allow the gases more time to become thoroughly ignited. They are sometimes formed of a brick wall, but oftener they compose a part of the boiler itself.

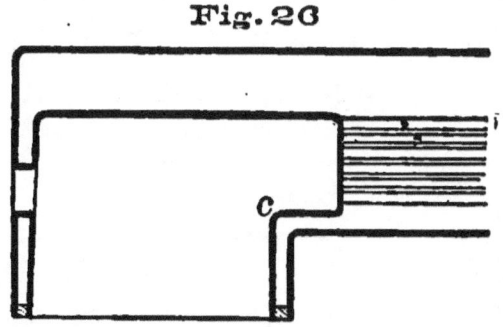

Fig. 26

Figure 26 shows the common practice in locomotive boilers. The intense heat playing upon the corner *c* produces so much steam underneath that the water is driven away, leaving the plate unprotected; and it rapidly burns out.

Figure 27 is from a marine boiler, and

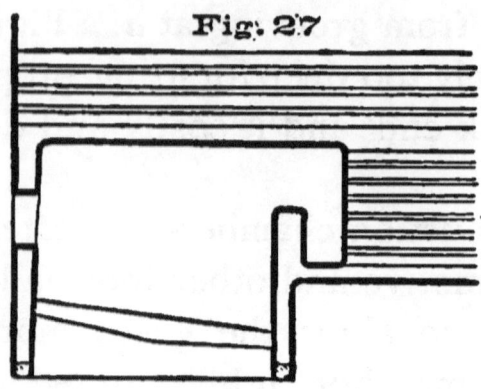

Fig. 27

is so dangerous that it is given here only to show how badly a boiler can be designed. Figure 28 gives a good arrangement for a locomotive boiler in lieu of an arch, which forces the heat too sharply against the crown sheet.

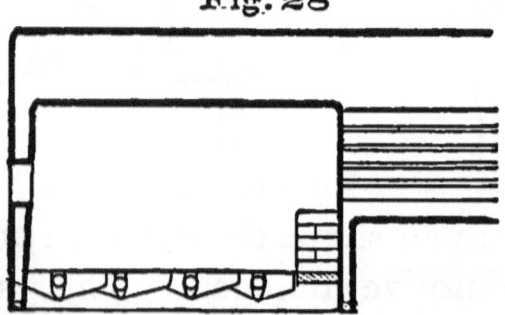

Fig. 28

By allowing a little space between the bricks and the tube sheet, the air which arises will materially aid the combustion.

The plates used for furnaces should be as thin as safety will allow, as heavy plates will blister or burn when exposed to the intense heat, whereas if the plate is thin the water will absorb and carry away the heat fast enough (other things being equal) to prevent such trouble. When a blister is once started there is no telling where it will end. As far as possible laps should be placed so that the currents of flame do not strike upon their edges. Care should be taken that there should be no places for the impurities of the water to lodge upon and bake. The water side of all lap-joints should be turned downward, and, as far as possible, above the water line.

IV.

RIVETED JOINTS.

Riveted joints are made either by lapping or by butting the plates. The lap joint is the cheapest, but at the same time the poorest, as it distorts what would otherwise be a regular form. It may be made passably good, however, by bending the plates in opposite directions so that the line of pull is directly through the centre of the lap and the faying surface as shown in figure 29.

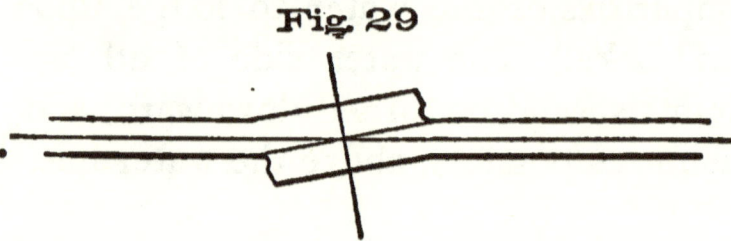

Fig. 29

This should be done whether the work is straight or cylindrical, except of course girth seams on cylinders where one course is made of larger diameter than its neighbor. Lap joints may have one or more rows of rivets, though seldom more than two. In cylindrical work the girth joints will be sufficiently strong with one row for all ordinary pressures, but the longitudinal joints should always be double riveted.

Butt joints are usually double riveted, as in figure 30, the welts being of

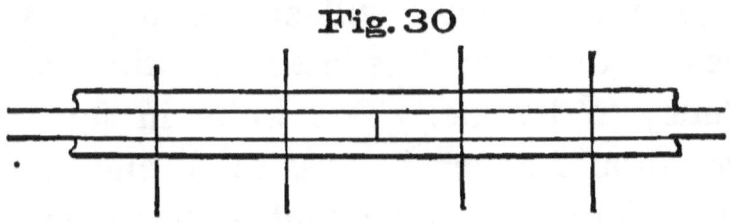

Fig. 30

equal width. A much better joint is that shown in figure 31, which has the

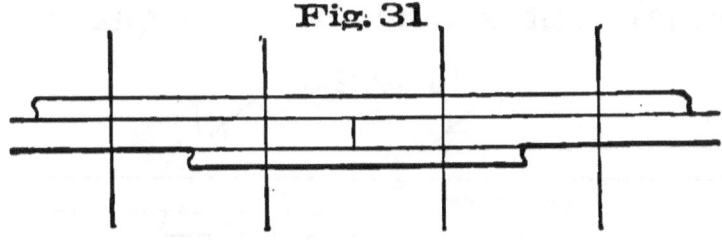

Fig. 31

outside welt double riveted, while the inside welt is narrower and takes but one row of rivets on each side the butt.

A very frequent fault in boilers is the grooving of the plates close against the edge of the lap on the inside, caused by the constant alternation of the expansion and contraction, that concentrates itself close to the stiffest part, which is the lap. This is also aided by bad calking. Now in figure 30 we have a treble stiffness just at the edge of the lap; whereas in figure 31, which is as strong a joint as the plate will stand, this stiffness is divided; thus lessening the tendency to buckle. The welts in figure 30 should never be less than one-half the thickness of the plates they cover, and are usually the next sixteenth of an inch thicker. In figure 31 the outside (wide) welt should be three-quarters

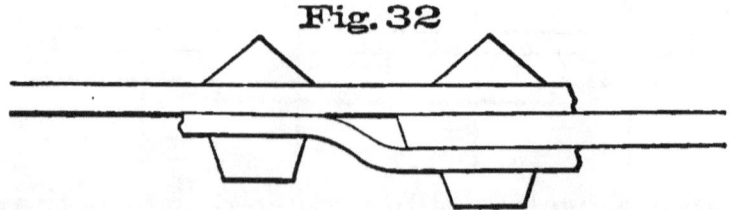

Fig. 32

and the inside one-half the thickness of the plates. The space between the rows of rivets should be about 25 per cent. more than that called for in the table, to allow for calking.

There are other styles of welted joints, the most usual being as in figures 32 and 33. Figure 32 is a vagary, though

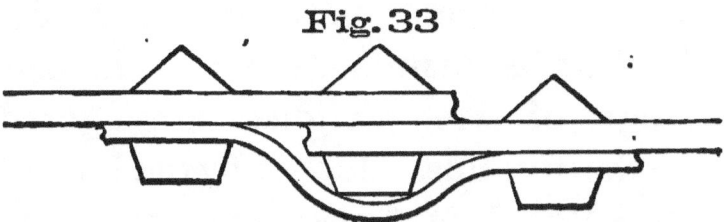

Fig. 33

often met with. It strengthens only one plate, while the other is as weak as ever, and the tendency to groove is greater than if the welt was not used. The welt in figure 33 is intended solely to prevent grooving, and for this purpose it is a comparative success. It is made of light iron, not over one-half the thickness of the plate, and is riveted with only one-half the number of rivets as the lap, if the lap is single riveted. If the lap is double riveted, use the same number as are in one row.

In the designing of riveted joints, the

point to be aimed at is to have the strength of the rivets and that part of the plates left between the rivet holes as nearly equal as possible. The proportions of riveted joints given in the accompanying table are compiled from the practice of several prominent and successful boiler manufacturers. The percentages of strength given are figured from the formulæ adopted by the United States Board of Supervisors of Boiler Inspectors. The excess of rivet strength, as shown, is made to allow for the almost unavoidable imperfections in driving the rivets. In the smaller sizes the excess is made greater on account of the difficulty of making the joint tight. If the idea of equal strength was carried out for these, there would be no end of trouble, as the plates would spring in the calking.

To find the strength of plate in the lap in per cent. of solid plate take:

$$100\frac{p-d}{p} = \text{per cent.} \dots\dots\dots 8$$

RIVETED JOINTS.

And to find the strength of the rivets in per cent. of solid plate take:

$$100 \frac{an}{pt} = \text{per cent} \dots \dots \dots \dots 9$$

Where p=pitch, d=diameter of hole, a=area of hole, n=number of rows, and t=thickness of plate.

PROPORTIONS OF RIVETED JOINTS.

Thickness of Plate.	Diameter of Rivet.	Length of Rivet.	Diameter of Hole.	Pitch — Single Row.	Pitch — Double Row.
¼	⅝	1¼	11-16	2	3
5-16	11-16	1½	¾	2 1-16	3⅛
⅜	¾	1¾	13-16	2⅛	3¼
7-16	13-16	2⅛	⅞	2 3-16	3⅜
½	⅞	2¼	15-16	2¼	3½
9-16	15-16	2½	1	2 5-16	3⅝
⅝	1	2¾	1 1-16	2⅜	3¾
11-16	1 1-16	3	1⅛	2½	3⅞
¾	1⅛	3¼	1 3-16	2 9-16	4
13-16	1 3-16	3½	1¼	2⅝	4⅛
⅞	1¼	3¾	1 5-16	2 11-16	4¼
15-16	1 5-16	4	1⅜	2¾	4⅜
1	1⅜	4¼	1 7-16	2⅞	4½
1	2	3	4	5	6

PROPORTIONS OF RIVITED JOINTS.

Distance apart the Rows.	Width of Lap.		Strength in per cent. of solid Plate.			
	Single Row.	Double Row.	Single Row. Plate.	Single Row. Rivets.	Double Row. Plate.	Double Row. Rivets.
1½	2	½	66	74	77	99
1 9-16	2⅛	3⅝	64	68	76	90
1⅝	2¼	3⅞	62	65	75	85
1 11-16	2⅜	4⅛	60	63	74	81
1¾	2⅝	4⅜	58	61	73	79
1 13-16	2⅝	4½	57	60	72	77
1⅞	2¾	4⅝	55	59	72	75
1 15-16	2⅞	4⅞	55	58	71	74
2	3	5	53	57	70	73
2 1-16	3⅛	5¼	52	57	70	73
2⅛	3¼	5⅜	51	57	69	73
2 3-16	3⅜	5⅝	50	57	69	72
2¼	3½	5¾	50	56	68	72
7	8	9	10	11	12	13

RIVETING.

Rivets are pointed in three different styles : conical, button and countersunk, of which the button and the conical when snapped are the strongest, from the fact that the snap compresses the metal, while in the hammered cone the metal is spread

Fig. 34

CONICAL   BUTTON   COUNTERSUNK

Cones are easier to get tight at first, yet if they should leak, snapped cones are just as readily calked as the hammered. Except for ornamental work, the snapped cone is practically as strong as the button which must be snapped. Cones should be kept pretty high, not less than three-quarters of the diameter of the

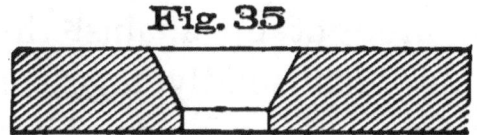

Fig. 35

rivet. If much lower, they will be weak.

The holes for countersunk rivets should be made at an angle of 60 degrees, but the taper must not go more than three-quarters of the way through the sheet. The countersinking tool should have a tit on its point to guide it and prevent eccentricity. A countersunk rivet should never be used where it can be done without, as it weakens the plate while its own strength is no greater.

Rivets are driven either by power or by hand. For tight work, the power driven are by far the best. Any power may be used, but to make a good job, it requires a double die or set; one to hold the plates together and the other to set the rivet. There are several machines in the market that do this to perfection. They have an annular die, which is brought up against the plates and holds them solidly together, while another die, working inside of it, sets the rivet. In such a machine, on girth seams of cylindrical work, do not start on one side and follow directly around, as the least slackness of the faying surfaces will be crowded between the last few rivets, making a bad spot to calk. But rivet across the quarters, keeping the joint well bolted, then take the eighths. This will divide up the slackness, so that it will not be troublesome. The rest can then be driven right along.

In power riveting, the rivet may be heated all over, as the head is held in a cupped die which prevents its spreading, and consequently the body can be upset its whole length, filling the hole solidly under the head as well as the point.

For hand riveting the head cannot be heated quite as hot as for power work, or it will spread too much under the holding-on hammer. This hammer should weigh about twenty pounds, and have a stout oaken or hickory handle from six to eight feet long. And, wherever practical, the hook, which carries it, should be about one-third of this distance from the head. This will allow a little play to the hammer, so that it will give a blow in answer to that on the point of the rivet, thereby aiding in upsetting under the head. For places where this hammer cannot be swung, a bar two and a half to three inches diameter and eighteen to twenty-four long can be used. And for narrow space, such as the sides of fire-boxes, which cannot be reached otherwise, a block of iron on the end of a long bar must be used. These last are not good for the rivet, especially if the rivet is hammered much after losing its heat. It becomes crystalized and is liable to break.

While waiting for a rivet, the plates should be well hammered together, drawing the bolt up tight in the next hole. Have no sharp corners on the hammers for this purpose, as they are apt to mar the plate.

If the holes in the plates do not match fairly, or shut by, do not use a drift. A drift always strains the plate, even if it does not start a crack. Ream them out until they are fair, and put in a larger rivet. A few of which should always be kept on hand to meet such cases. For reaming holes a good tool is shown in figure 36, *a* being the cutting-edge. The section is enlarged to give a better idea of its shape. This tool is stronger, and will cut faster, cleaner and

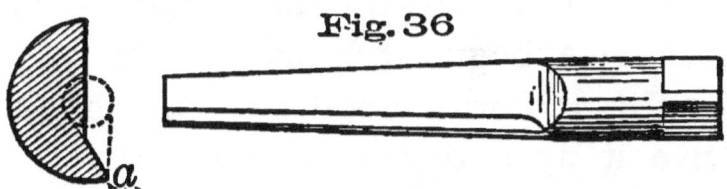

Fig. 36

easier, than the usual half (oftener less than half) round ones.

When the helper passes the rivet through, it should be held by pressing lightly against its side, until he has his holder in position. Do not strike it sideways, or the chances are that the point will be lopsided, and if it should happen to leak, the only remedy is to cut it out. Work lively now, striking the rivet square on the end until it begins to spread over the plate. Then, as fast as may be, shape the point. It should always be borne in mind that for good work it is just as important to fill the hole up solidly as to make a neat point, and this cannot be done without quick and hard upsetting.

For snap work, the upsetting is done in the same manner as for hammered points. But instead of hammering over as the rivet begins to spread, set the snap over it and with good swinging blows drive it up until the snap just touches the plate. The hammer for this work should be at least twice as heavy as the snap If no heavier, the snap itself will absorb so much of the blow that its effect on the rivet will be very feeble. The lip of the snap should be slightly rounded, for, if sharp, it is liable to cut

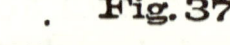

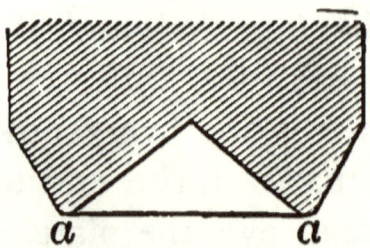

the plate, thereby further weakening it in its weakest part. See *a a* in figure 37.

If the rivet holes are drilled with the plates together, the plates should be taken apart and the burrs taken off before riveting up. The holes should be slightly countersunk on the outside as well.

The shop should never be allowed to get too cold in winter, for the sake of the work as well as for the sake of the men. Because, if the plates get too cold and the rivets are driven as well as they should be, when the boiler begins to warm up quite a number of rivet heads fly off. They often break at the fay, giving no end of trouble. Others still will simply crack partly through. Of course the users of the boiler blame the maker; the maker, the manufacturer of the rivets ; and the manufacturer of course retorts that the rivets were burned in the heating. Yet all the while the trouble was that the boiler was too cold. Sometimes the heads drop off before the boiler is warmed up. This seldom occurs, however, except in case of very heavy plates or an extra number of plates are brought together.

V.
BRACING AND STAYING.

All shapes not *complete* circles must be thoroughly braced or stayed. This includes all elliptic and partially circular forms as well as flat.

The term "brace" is applied to long rods reaching across the boiler, while "stay" is applied to short bolts and other devices in corners or on one side only.

The first and most important point is to have a sufficient amount in number and size, and if the plate is flat or nearly so, they should support it entirely without regard to the plate's own stiffness. The second is, to have them so disposed as to present the least obstruction to a free inspection. And last but by no means to be lost sight of, to allow a free circulation of the water.

BRACING AND STAYING.

Wherever possible, braces and stays should draw at right angles to the plate that they are intended to support. If this cannot be done with a direct brace and without interfering with the accessibility, a modification of the brace or a stiffening bar braced at another point should be used. It should never be necessary to remove a brace to reach any portion of the interior. If a brace is removed from any cause, the chances are equal that it will not be replaced.

Braces should be fitted to their places tight enough to prevent shaking, and each brace should draw just as much as and no more than its neighbor. Care should be taken not to overload a brace by straining it in, as it has quite enough labor to perform in service.

Too much care cannot be taken in fitting stays and braces. Being out of sight and for considerable periods of time, especially in case of new boilers, there is no way of determining their efficiency, save by the failure of the boiler, which generally occurs at some critical time.

The simplest form of stay is a plain rod threaded its whole length, screwed through two opposite plates, and having

Fig. 38

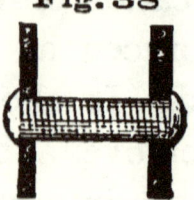

its ends riveted over; figure 38. It is only used in narrow water spaces and should be arranged in vertical and horizontal rows so that bars and scrapers may be more readily passed between them. The thread should be fine enough to have not less than three full turns in the plate. Four is better. The usual practice in locomotive fire-boxes is twelve per inch, the bolt being $\frac{7}{8}''$ diameter. Now with a tensile strength of 50,000 pounds per square inch this bolt will support 50,000 times the area in square inches at the base of the thread, or $\dfrac{50,000 \times .7765^2 \times .7854}{6} = 3845$

BRACING AND STAYING. 61

pounds, using 6 for the factor of safety. .7765 is the diameter at the base, the thread being of the Franklin Institute Standard. To find the distance apart, divide the 3845 by the working pressure required, say 140, and extract the square root. Then $3845 \div 140 = 27.46$ (the total area the bolt can support) and $\sqrt{27.46} = 5.24$. Another element to consider in this connection is that of corrosion, and although the bolt may be amply strong when new, it is better to err upon the strong side. On this account as well as faults of workmanship, which it is impossible to entirely avoid, it is customary to place them not over $4\frac{1}{2}''$ apart.

The distance apart of the bolts allowable from the thickness of the plates, can be found from the following formula:

$$p = \frac{2t}{\frac{\sqrt{PK}}{c}} \dots\dots\dots\dots\dots\dots 10$$

In which $p=$ the pitch in inches.

$t=$ thickness of plate in inches.

$P=$ pressure per square inch.

$K=$ factor of safety $=6$.

$c=$ tensile strength of plate $= 60,000$ pounds.

Assuming t to be $5\text{-}16''$ which is the usual thickness of locomotive fire-boxes, we have $(.3125 \times 2) \div \left(\sqrt{\dfrac{140 \times 6}{60,000}}\right) = 5.28$ or about the same that the bolt will support. The same trouble from corrosion occurs with the plate, coupled with the wear from the fire. They rapidly thin down the plate so that $4\frac{1}{2}''$ pitch is none too close.

This style of bolt should be fitted very tightly especially in a steel fire-box. It requires as great a force to screw it in as it requires to tap the holes. The tap for this purpose is best made with the reaming portion long enough to run clear through both plates before the thread begins to cut. The thread, one-half of which may be a half-thread deep, should be long enough to reach the same distance. The shank also should be as long, so that the tool may be sent clear through and not have to be backed out. Backing out is apt to strip the thread in steel plate. The appearance

Fig. 39

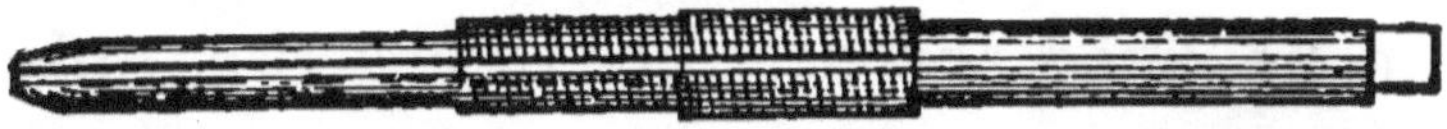

of such a tap will be as in figure 39 which is drawn from one of three, which together tapped over fifteen thousand holes through ½" and ⅜" plates with 3½" to 4½" space between, and no one of the bolts have failed from faulty tapping after nearly three years continuous service. The diameter of the reamer part should be neither more nor less than the diameter of the bottom of the thread. If larger, there will be a pinhole leak, and if smaller, the tap will almost invariably strip the thread, if the plate is steel. Too much importance cannot be given to these two points, as well as to fitting the bolts tightly in the thread.

Where facilities for cutting off the proper lengths of these bolts from long rods, cut in a bolt machine, the very handy socket wrench shown in figure 40

Fig. 40

can be used for setting them, the hole being tapped the depth required for the projection of the bolt for riveting.

The wrench for screwing in these bolts should be double handled, from 30" to 36" long, with the ends turned up 4". There should also be several single handled ones, on hand for use in concave parts such as where the straight part of a locomotive fire-box swells out to the barrel.

A projection of one-half the diameter of the bolt should be allowed on each end for riveting. Upset square on the ends and snap with a shallow tool, the object being not so much to cover the plate as to fill the hole solidly. Hammering down the edges has the effect of

BRACING AND STAYING. 65

breaking the thread in the plate and in the direction of the strain as well.

Several of the upper rows of these bolts are frequently made hollow, with the inside end plugged. Or they may be drilled in their outer ends only. In such cases it is advisable to have them, ⅛" at least, larger in diameter to make up for the loss of strength. The hole may then be of any diameter up to 3-16" diameter. The steam blowing through the hole serves to notify that the bolt is broken. This would not be likely to happen if the fire-box was properly stayed at the top.

In locomotive boilers with straight top and arched crown sheet as in figure

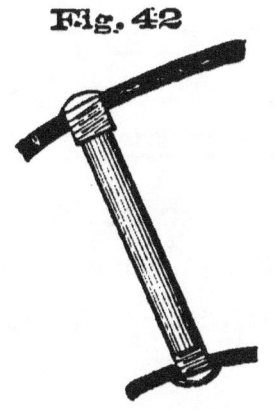

Fig. 42

41 this bolt is used with a modification, figure 42. One end is enlarged ⅛" in

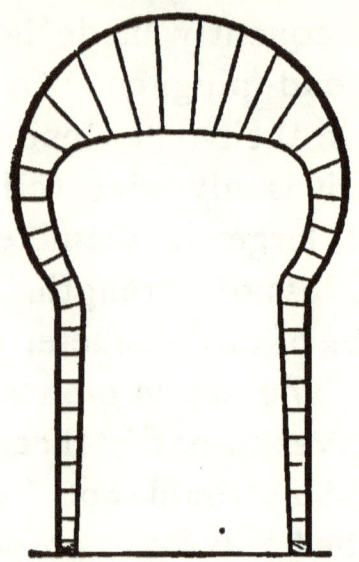

Fig. 41

diameter to allow the greater part of the bolt to pass through without screwing. This necessitates a tap like figure 43

Fig. 43

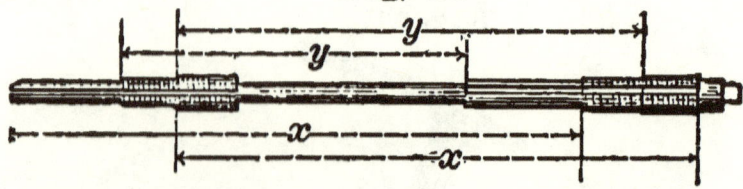

The dimensions x to be long enough to take the longest distance over the plates and y the shortest distance between them.

Figure 44 is another type of bolt for

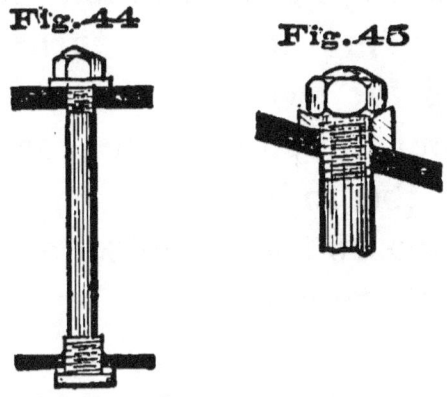

the same purpose. This requires a copper washer, which should be not less than $\frac{1}{8}''$ thick, under head and nut. Where this bolt is used at an angle to the plates the head and nut must be faced off to a cone shape, and the copper washer made to fit, as in figure 45. Figure 44 is frequently used in locomotive boilers for staying the heads, running from end to end, and under the crown bar. In this case the bolt is not screwed into the front head, but has a nut and copper washer on the inside as well as outside. It is essential to have a support for these bolts about midlength, which may be a plain bar, of 4″ × 1″ iron, with holes, through which the bolts run, and having its ends turned at right angles for riveting to the boiler shell. Without this support the continual surging of the rods, when in service, would soon loosen them and set at defiance all attempts to make them tight. Properly fitted there is no better stay for that portion of the back head opposite and a little above the crown sheet. A tee or angle iron with a diagonal brace is difficult to get in, the space being so limited. In some cases this place has been supported (?) by a bolt like figure 1, screwed into the crown bar, an arrangement that can not be too heartily condemned. The duty of a crown bar is severe enough without subjecting it to any other than its legitimate strains.

Various styles of bolts are used for fastening the crown bars to the sheet, including a plain rivet with head above; a common bolt with end tapped through and riveted over the sheet; the same tapped through the sheet but nutted on the fire side. Figure 44 is also used. A nut should never be used in the fire, as it is liable to fail at any moment. Nor is it advisable to tap through the crown sheet. In case of leakage or failure of any kind, the bolt will require replacing and the hole must needs be retapped a little larger. It frequently happens, as well, that the water contains such impurites that it necessitates the entire removal of the crown bars for cleaning. It will readily be seen what trouble and expense arises from riveting or screwing into the sheet. The best style of bolt used at present is shown with its washer in figure 46. This bolt is now kept in stock by several bolt manufacturers. The difference in cost over a plain bolt is more than offset by the saving in tapping. A bolt of this kind is tapered under the head for about an inch of its length, increasing its diameter 1-16"; so that if the hole in the plate is 1-32" larger than the body of the bolt this tapering portion may be driven in solidly. The chips or shaving caused by the driving fall into the little groove under the head, and therefore do not require the withdrawal of the bolt for clearing out, so that if once driven solidly it remains there. The

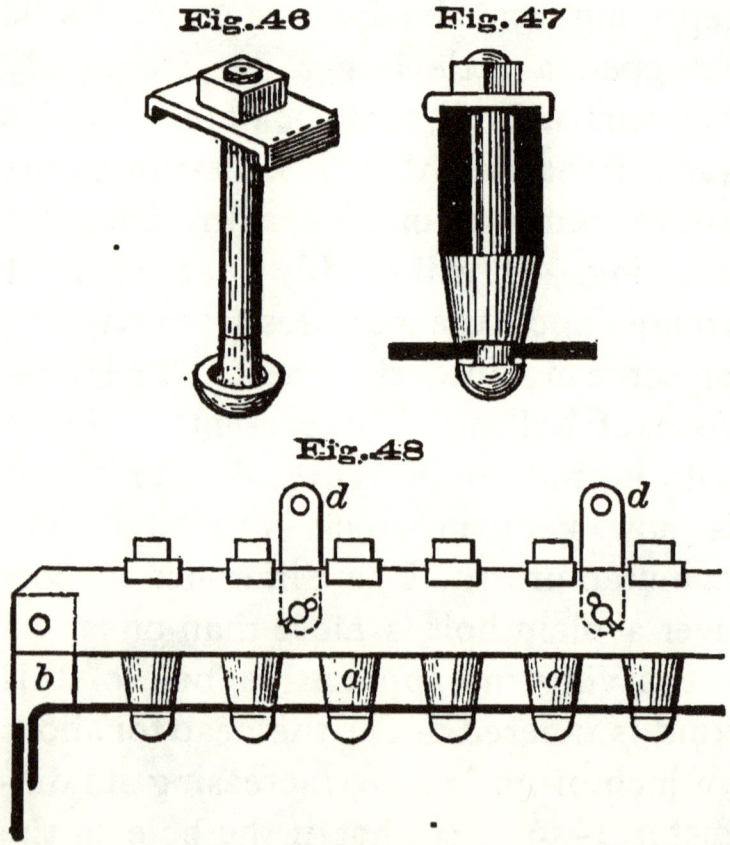

rule for spacing crown bar bolts is the

BRACING AND STAYING. 71

same as for ordinary stay-bolts. This also determines the distance the bars will be from centre to centre. The style of crown bar most frequently used in this country is shown in section in figure 47, and in side elevation in figure 48. The thimbles *a* and the toes *b* are of cast-iron, but often the thimbles are of wrought iron—a needless expense—and the bars are welded together at the ends

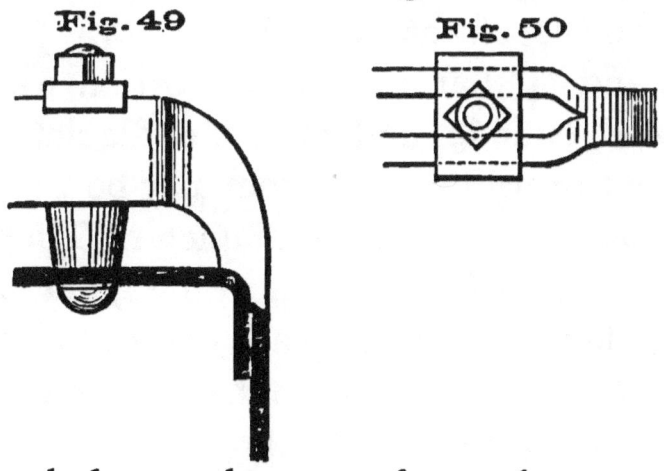

Fig. 49   Fig. 50

and drawn down to form the toes, as in figures 49 and 50. The welded toe has the very desirable abvantage of covering less of the plate than the cast-iron ones. It also allows a better circulation of the water. The toes should be closely fitted to the edge of the side plate and the turn of flange of the crown sheet. Be careful to have them long enough so that they will not be raised from their bearings when the thimbles are in place. The tap should be caulked (on the outside only) before the bars are fitted. The thimbles should be tapering so that while giving a bearing the full width of the bar they will present as small a surface to the bar as possible. With a $\frac{7}{8}''$ bolt and $\frac{3}{4}''$ bar, a good proportion will be $2\frac{1}{2}''$ outside and $1''$ inside diameter at the top, while the bottom may be $1\frac{1}{2}''$ outside and $1\frac{1}{8}''$ inside diameter. They should never be less than $2''$ high, and if long stay rods run beneath them $2\frac{1}{2}''$ is none too much. There being such a small amount of water above the crown sheet—gauge-cocks, etc., being set usually to carry but $3''$ to $9''$ depth—that with the great heat playing underneath, the steam forming below the bars has a great tendency to blow the water therefrom with frequently disastrous results.

BRACING AND STAYING.

The size of these bars may be determined from the formula for a beam supported at both ends and weight equally distributed, though the practice for such bars is $4\frac{1}{2}'' \times \frac{3}{4}''$ for crown sheets $40''$ to $48''$ wide. With the above thickness of bar, and disregarding the crown sheet which is of very great assistance in such a construction, to find the height make

$$h = \frac{\sqrt{w \times k \times l}}{8c \times b} \dots\dots\dots\dots\dots\dots 11$$

in which $h =$ height in inches.

         $w =$ total load in pounds.

         $k =$ factor of safety $= 6$.

         $l =$ length of span in inches.

         $c =$ tensile strength of bar in pounds.

         $b =$ total breadth (thickness) of bars in inches.

With a span of $44''$ and a pressure of 140 pounds per square inch and the distance centre to centre of bars $4\frac{1}{2}''$ we find $w = 44 \times 4\frac{1}{2} \times 140 = 27,720$. Then we have $(27,720 \times 6 \times 44 = 7,318,080) \div$

($8 \times 30,000 \times 1\frac{1}{2} = 360,000$) $= 20.328$, the square root of which $= 4.5$, the height required.

In figure 48, *dd*, are short links to which the slings are attached, and in this connection it is well to consider the enormous force which these slings are required to resist. It is an amount equal to the diameter of shell × length of fire-box shell × pressure per square inch less that amount resisted by the back head and throat plate. All of the remainder not entirely resisted by the slings must needs be thrown upon the short stay-bolts on the sides. That it is not carried by the slings in many cases is evidenced by the fact that so many fire-boxes are fitted with hollow stay-bolts near the top, to give notice of their breaking, which is from no other reason than lack of slings in number or size. To determine just how much strain will come upon the slings is difficult, if not impossible, but two slings having a sectional area of from 1 to 1¼ square inches to each alternate bar, are not enough. There should be at least two slings to every bar, and of nearly. if not twice such area.

The slings are made in three parts; the foot, the brace, and the links. Sometimes the brace and foot are in one. This is objectionable, as it necessitates cutting out rivets to allow removal. The link *d*, figure 48, is a flat bar, with a hole at each end, just long enough to allow the fork of the brace to be made fast to it, while the lower pin is in the center of the bar. These links are sometimes forged to a shape like figure 51. A

Fig. 51

square rod of proper length welded in hoop form is flattened down with pins to keep the eyes open. This makes a very neat job, but it is liable to straighten in service, unless it is welded together, which makes it expensive. The foot has a plate through which it is riveted to the shell. The lug is welded to this plate, and this should be done with the greatest of care. A good way is to mortise the plate, and drive the lug through from the back. Figure 52. The lug being upset it can be driven in

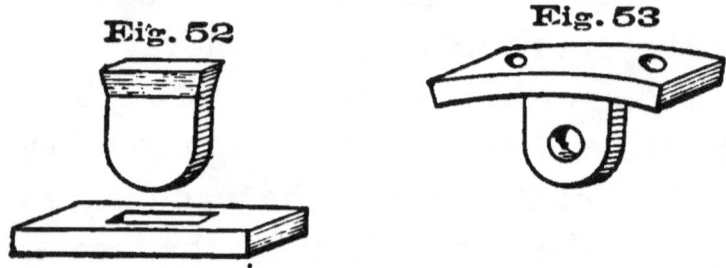

solid, and finished off at one heat with very little trouble, the finished foot appearing like figure 53. Where the brace comes into a dome or upon a side surface, it is made like either figure 54 or 55. Where the twist is not too short, nor the offset very great, figure 55 is the best, as it requires no welding. Figure 54 is stiff, and if the welding is thoroughly done, it is naturally superior. But welding in such work can seldom be guaranteed.

BRACING AND STAYING. 77

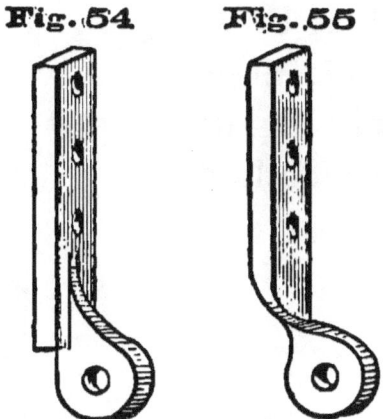

Fig. 54 Fig. 55

The brace itself may be made of a rod with jaws on both ends like figure 56. For such short braces as crown

Fig. 56

slings, two flat bars slightly riveted together, with washers between them, are much to be preferred, as they are not welded, and can so readily be marked off in their places. Take one of the bars and drill one end. Then set up with pin in place, upper end is best, hold the other end against the link, and scribe through the hole. This bar can then be laid upon the other, and used as a jig to drill it by. Figure 57 shows a brace of this description, which commends itself for its cheapness, as well as for its reliability.

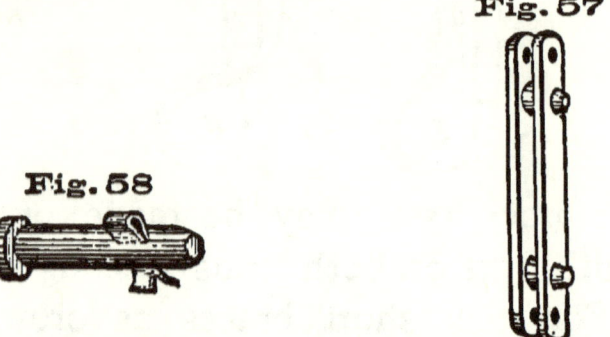

Fig. 57

Fig. 58

The pins for jointing braces should be parallel, and *never* be split to retain them in place. A nut gives great trouble in removing. After a short time only it may necessitate chipping off. This can be avoided by using a split cotter. Figure 58 is a pin that gives good satisfaction, and can be made entirely by machinery. Parallel in body, with point tapered so as to be entered easily, the cotter preventing its working out, it does not bind the parts together, so that if the brace does not draw, it is

BRACING AND STAYING.   79

easily detected. This cannot be done with a nutted bolt.

With the crown bar we have just been considering, using two slings, we have 27,720÷2=13,860 pounds, for each sling to carry. Then with a tensile strength of 30,000 pounds per square inch of section, we must have 13,860÷30,000= .462 of a square inch without allowance for a safety factor which would make it .462×6=2.772 or over $2\frac{3}{4}$ square inches. This, considering the support given by the end plates of the fire-box, may be excessive, and a factor of 4 may safely be used, making it .462×4=1.848, or about $1\frac{7}{8}$ square inches. If better iron than refined is used, then their size may be reduced in proportion, substituting the tensile strength for the 30,000 in the calculation. It may be well to note that not a few boilers are now in use, having but two slings to each alternate bar, and the section in the slings but little over $1\frac{1}{4}$ square inches. All are more or less troubled with leaky and broken stay-bolts. The only thing that saves the boiler from destruction is the heavy leg frame and the heads, and not the inherent strength of the construction.

Figure 59 is a plain crowfoot brace,

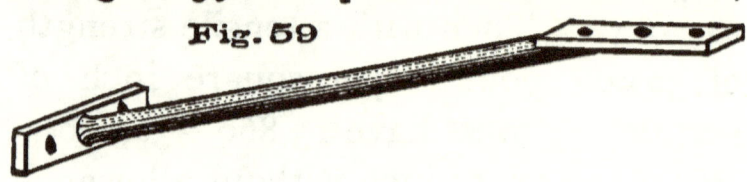
Fig. 59

which is very handy and good where a short stay is desired. Do not be afraid to put large rivets in the feet, as some boiler-makers appear to be, often using two $\frac{5}{8}''$ rivets to hold a rod $1\frac{1}{8}''$ diameter. A comparison of the areas will at once show that the rivets are too small.

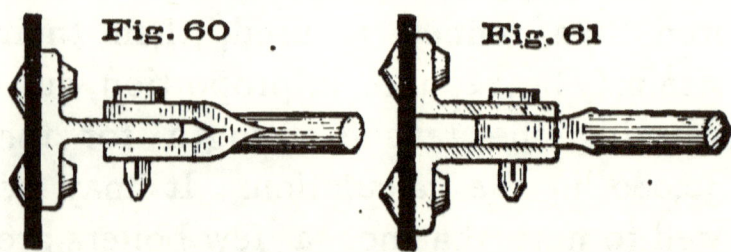
Fig. 60   Fig. 61

For long braces, tee and angle iron is used for feet, with an eye on the rod to fit over or between as in figures 60 and

61. The angle iron in figure 61, is objectionable for its tendency to straighten, and thereby acting as a powerful lever to pry the heads of the rivets.

Figure 60 is good as long as the surface to which it is riveted is flat, but if it is curved, it is dangerous from the fact that rolled tee iron is never solid at the root of the stem. This will show for itself very plainly if an attempt is made to bend it, hot or cold. The best way to make a tee is to flange it up from a flat piece of Ex. Flange iron or of steel, to a section shown in figure 62. This tee iron can

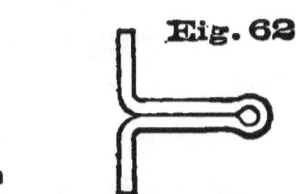

Fig. 62

be curved readily both ways, and is the only thing that should be used at the sides of locomotive "wagon top" boiler just forward of the fire-box. The method of forming this tee will be given under the head of smithing.

A few rivets through the stem of the tee close down to the corner will enhance the stiffness to such an extent that there will be but very little difference, if any, between it and a solid rolled bar. In placing such rivets be careful to keep them away from the spaces required for the jaws of the braces.

It may be, where the shape of the boiler is particularly crooked, an advantage to make the tee of two pieces of angle-iron, so as to fit separately. In such a case they must be riveted together after fitting. Drill or punch the holes in one, and lay together in their places, and mark the other from it. After riveting together, try it in and refit, as there may be a little alteration in shape caused by the "drawing" of the rivets.

Lugs for single braces can be made up in a similar shape, but as they would be narrow they should be welded in the stem.

Lay out the rivets for such tee irons by the same rule as used for stay-bolts on the sides of fire-boxes. That is, with due regard to the diameter of rivets used.

Simply stiffening a surface with tee or angle iron without bracing from an opposite point, should not, under any consideration, be allowed. The effect of so doing is to throw such weaving ing strains upon the plate at the turn of the flanges, as to soon crack them through. Two boilers that were built or repaired (?) by a prominent railroad corporation, and whose mechanical engineers are held up for patterns of expertness, came recently under the author's observation. The back heads had to be replaced by new ones. Both were badly cracked on each side, a a figure 63. Patches had been applied on both sides of one, and one side of the other, and two of the patches even were cracked. The tee was made of two $6\frac{1}{2}'' \times 4'' \times \frac{1}{2}''$ angle iron, the long

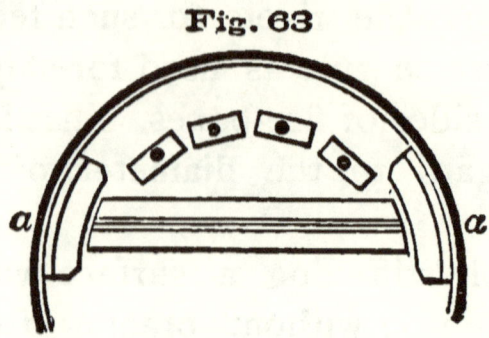

Fig. 63

legs riveted together, and reaching as close to the sides as the curve of the flange would permit. Good enough in itself, but being so stiff that it concentrated all the spring upon the flange. Yet had it been reinforced by a few stay rods, to some opposite part, there would have been no trouble whatever.

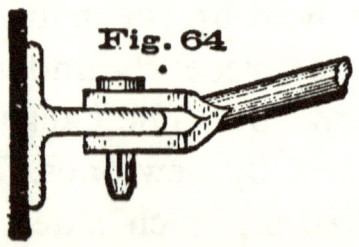

Fig. 64

In laying out for tee iron brace feet, be careful to so dispose them in such directions that the fork of the rod does not meet them at a transverse angle, as in figure 64. The evil effects can be seen at once from the figure. They may be swung around the axis of the pin without any bad resuts, however.

The opposite ends of these braces may be made fast to feet like figures 55 and 56. On account of removal, these feet should never be forged on the rod, although it is frequently done.

In large boilers to reduce the *number* of rods, recourse is frequently had to such devices as are shown in figures 65 and 66, care being taken to make the rods larger to correspond.

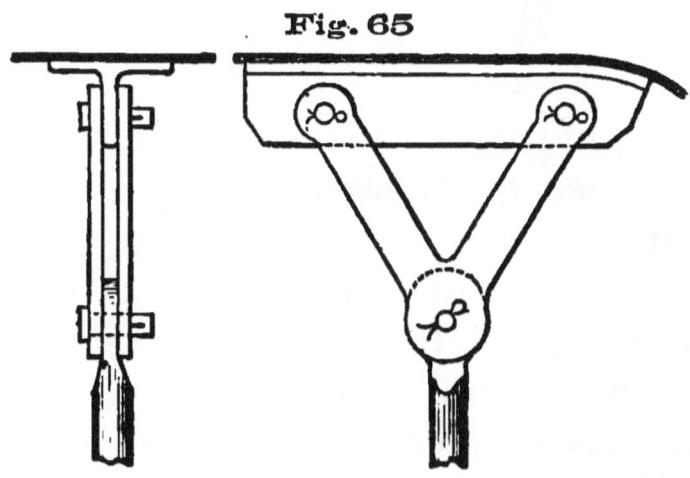

Fig. 65

86　　　BOILER MAKING.

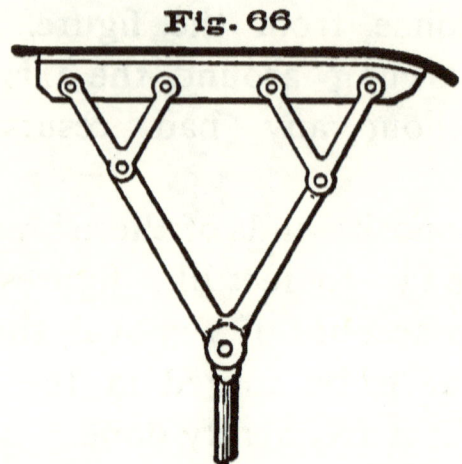

Fig. 66

Figure 67 is a modification, often

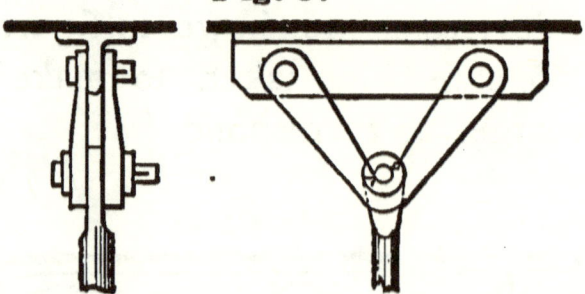

Fig. 67

used, of figure 65, and it is claimed to be cheaper. It saves drilling one hole, but it would seem to be much more troublesome to hang on account of the dropping of the ends. To set these heavy braces requires heavy turnbuckles. An example of this is given in figure 68.

BRACING AND STAYING.   87

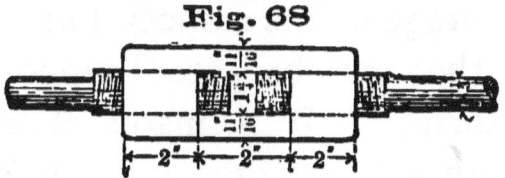

Throat or belly stays in locomotive boilers, from their being exposed to the sharpest current of flame or hot gases, must be thoroughly made and should cover as little as possible of the surface of the plate that they support.

Figure 69 is largely used for this pur-

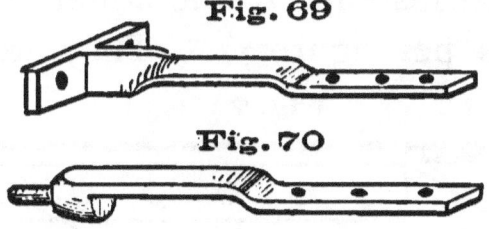

pose being a simple crowfoot brace with the stem made of flat bar. Usually of 2"×⅞" stock. The foot 7" long with the rivets 4½" centers. Being riveted to the throat sheet it is necessary to offset the stem to pass over the rivets of the lap. Some master mechanics reject this stay as covering too much surface, and use a brace like figure 70, which has a decided advantage in this respect, but requires twice the number to support the same surface. It is made of 2"× ⅞" iron with a lug forged on the end in which to screw a plain stay-bolt. The end of this stay is kept clear of the tube plate at least ½". An inch would be better. The stay-bolt screws through the sheet as well as the lug and is riveted over upon the inside of the fire-box.

Gusset stays are those made of plate iron and fastened to the boiler by angle iron, as per figure 71. They are prop-

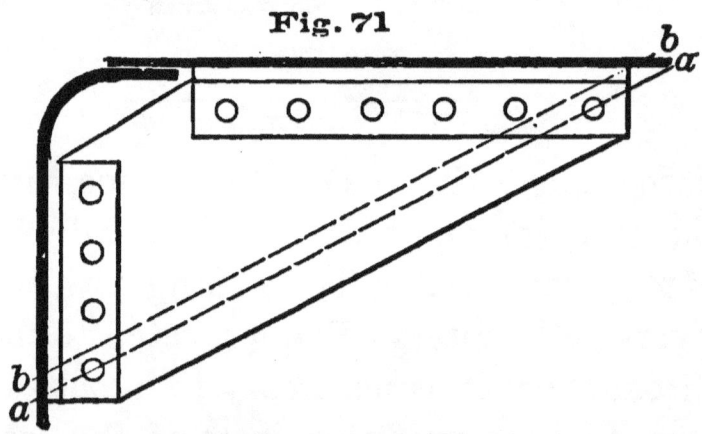

Fig. 71

erly set radially to the boiler so that the angle iron fastening them to the shell will then require the least amount of smithing. The objection usually given to these stays is that the greatest amount or strain is thrown immediately upon the edge of the plate, and on this account most engineers refuse them. This is something of a mistake, as the greatest strain does not come upon the edge, but upon a line drawn through the centre of the outside rivets, *a a*, in the figure. It is evident, then, that the width from the edge to a line (*b b*) drawn in the centre between the first and second rivets will be the amount to resist the strain. In a stay in actual use this width was 3½″ and the plate ⅜″ thick, making a sectional area of 3½ × ⅜ = 1,3125, equivalent to a round rod over 1¼″ diameter, surely sufficient to support any surface that the rivets would hold, without counting the support given by the remainder of the plate. A great advantage of such stays over long rod braces, is their leaving so much space for working in.

Figures 72, 73 and 74 are familiar

90   BOILER MAKING.

types as used in marine boilers. Figure 72 is bad in that it covers too much of the surface of a heated plate, and that it depends entirely upon its own stiffness to resist the strains. Figure 73 is much better, yet the crowfeet cover too much surface. Figure 74 is the

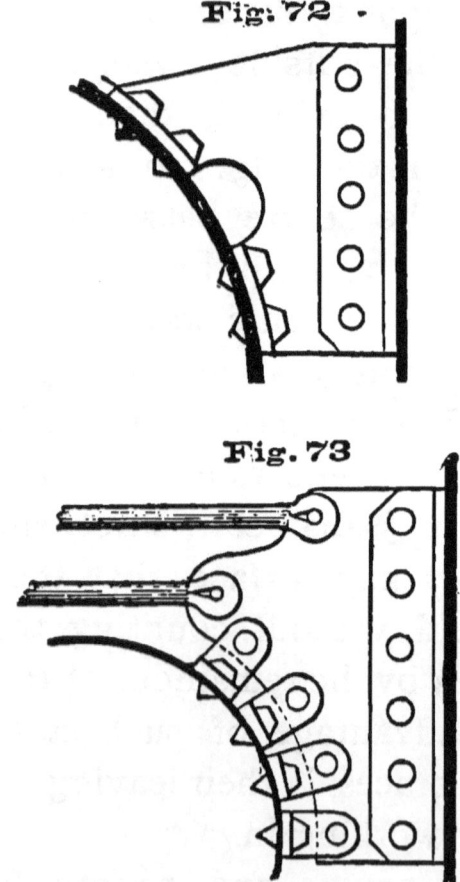

BRACING AND STAYING.

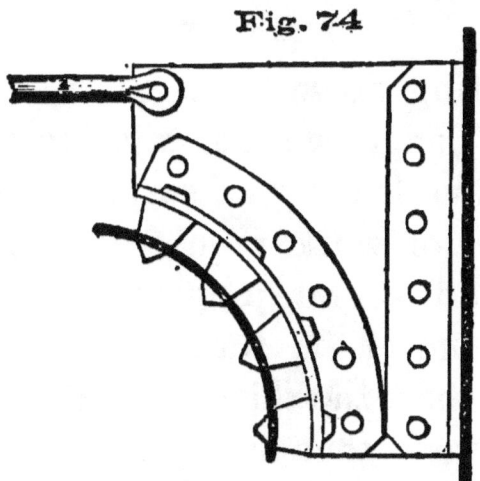

Fig. 74

best of the three, being stiff, strong, and presenting the least amount of surface to the plate. Never use a brace or stay that is bent over to form a foot on one side only. Being liable to straighten out, no reliance whatever can be placed upon them.

VI.

FLANGING.

Bending down the edge of a plate to form a flange, without injuring it, at the same time leaving its surface smooth, is a piece of work requiring skill, quickness and sound judgment. A flange should never be turned of less inside radius than the thickness of the plate; nor greater than sufficient to allow the same distance from inside to centre of rivet line, as from centre to centre of stay-bolts required to support the flat portion of the same plate. Up to this limit, the greater the radius the better, as the fibres of the metal are not so severely strained nor is the plate so much reduced in thickness.

There are two general methods of flanging. One with a "former" large enough to shape the whole plate over; the other with but a small iron block or anvil. Sometimes even a block of wood is used. For thoroughly satisfactory work the first plan only can be used. And only lack of time or other compelling cause should allow of any other method. The cost of casting the full former is practically nothing as compared to the cost of time and fuel saved. The iron is always good stock, and can be broken up and remelted if it is not thought desirable to save.

The former should be heavy, say $1\frac{1}{2}$ thick, with a flange around it of the same depth as the flange desired on the plate. The pattern should be made with a standard rule, no allowance whatever being made for shrinkage. In such castings the shrinkage is but very trifling, and the flange, no matter how closely it is driven down, will always spring off sufficiently to make up for what little occurs.

The opening for the female flange of firedoors on the same account should be made larger than the standard rule will give it. Not less than 3-16″ per foot, while that for the male flange should be from the standard. This to allow some freedom when setting up. When together, the male flange is heated and expanded with mauls to fit. The flanges of the former should be thick enough to allow one jaw of a clamp to be set on the edge of it, while the opposite jaw is resting on the flat of the plate, and clear of the turn. If the turn is of large radius there may, to save weight, be a supplementary flange or rib, parallel to the outside flange and inward a sufficient distance for the purpose. Figure 75 shows a section of former with plate clamped upon it.

Fig. 75.

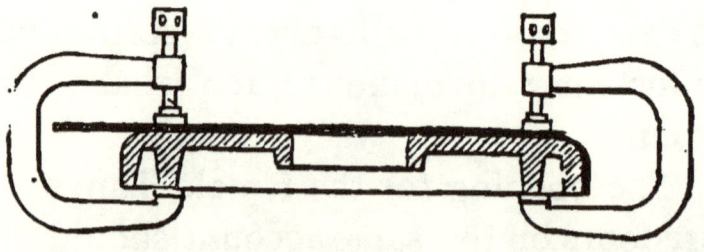

Having a former of the full size and

proper shape, no laying out is necessary until after the plate is flanged, except to locate a couple of holes for the steady pins, which hold the plate in position while working the flange. These holes should be made of suitable size for closing with rivets. Be careful to place them so that they will not interfere with the stay-bolts. Corresponding holes are required in the former. The pin should be somewhat tapering, so that it can be driven in tightly. It is easily loosened by rapping the plate close beside it.

The former, especially about the turn of the flange, should be examined for roughnesses and irregularities, and all carefully smoothed off.

The fire for flanging should be long and narrow, the idea being to heat as long a section of the plate as possible, without going far back from the line of turn. Making the tuyere of a pipe with a row of holes longitudinally, will enable this to be done. If much circular or curved work is to be done it will pay to have a curved tuyere.

Have a good coke fire and bank thoroughly with wet coal or fire brick. After laying the plate on, bank carefully all around it so as to confine the heat to the plate and not waste it and time by blowing it away. Bring up the heat *slowly*, especially for steel. For iron a good bright heat is wanted, but for steel a light cherry red is as much as it will stand.

Lose no time between the fire and the former, for the plate rapidly loses its heat, and it should never be necessary to heat twice for the same section. Three bail hooks hung from a scale beam lever, a "monkey tail" at the opposite end, the whole carried on an overhead traveller, is a very handy device for carrying the plate back and forth. See figure 76.

The forge should be low; not over 18" from the floor. The former should be blocked up about a foot high, to

FLANGING. 97

allow freedom for entering the clamps.

When the heat is right, get your hooks under the edge of the plate

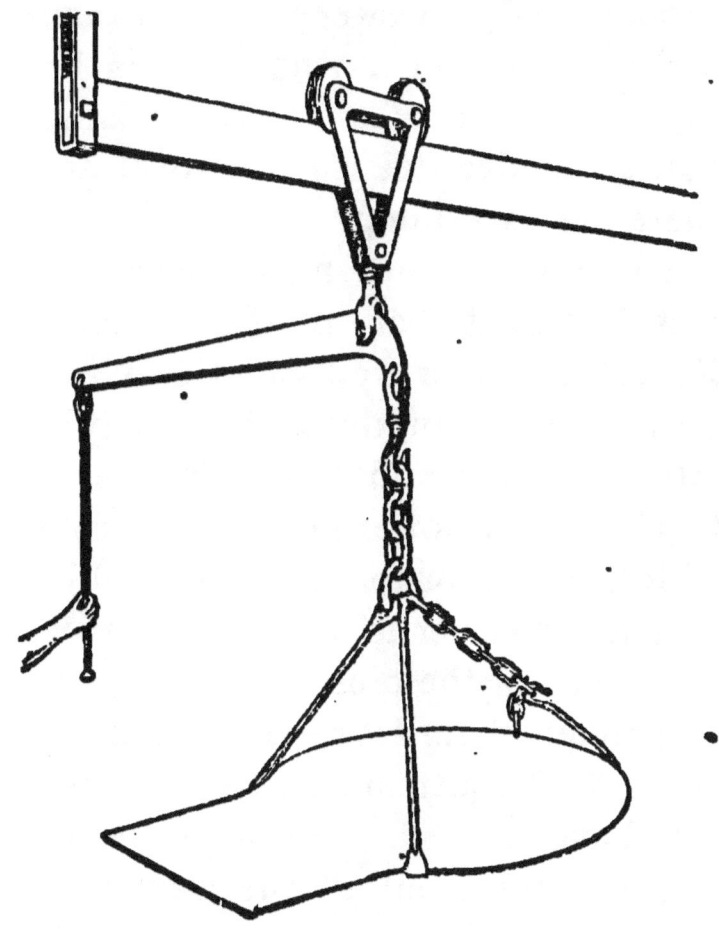

Fig. 76.

before you clear off the top. While doing this reduce the blast, but do not shut it off entirely until the plate leaves the fire. In other words, keep the heat in the plate as long as possible, at the same time working over the fire without being choked with the gases. Have a broom ready to sweep off the cinder, and dust as soon as the plate is lifted off. Do not be afraid to stoop and see that the under side is perfectly clean. If there are hard lumps burnt on, so that the broom will not move them, use a chisel-pointed bar to cut them off. Keep the former well swept also.

If all is clean, drop the steady-pins into the plate and swing it over the block, gently lowering it while fishing with the pins for the holes in the block. It is much easier to find the holes with the points of the pins while the plate is suspended, than if it were down so that it could not be seen under. As soon as the pins have entered, let go with the hooks and run them out of the way. Drive the pins in solidly, and set several clamps along the edges away from the heat. One to be close to each end of the heat.

FLANGING. 99

To handle plates about or steady them while hanging, a porter hook is a deal more pleasant and effective than leather mittens. There should be a straight and a right angled one for every man at work about the flange fire. The mouth should be about 4" deep and wide enough to take the thickest plate used. Jaws 1" thick and 1½" wide at widest part. Handle about 1" in diameter and 24" to 30" long. Figures 77 and 78 show the two kinds and the method of using.

Fig. 77.

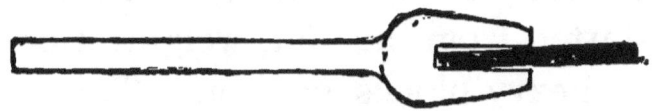

(See fig. 75, page 94),

Fig. 78.

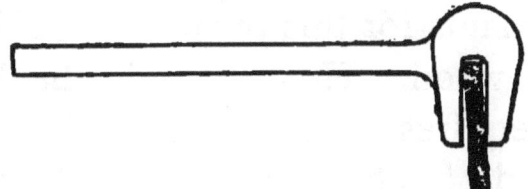

Figure 78 can be used either

side up with equal facility. These hooks are handy anywhere in the boiler shop for either plate or angle iron. In carrying materials with them the men can face the direction in which they are walking, thereby making better progress, and with less danger of stumbling.

As soon as the plate is secured to the former, beat down as fast as possible with round faced mauls, striking at first about the middle of the overhang. If struck too near the edge the plate will buckle up, and if too near the flat the edge will turn up. As soon as the turn is started from the flat, weight it down with heavy blocks of iron. This prevents the flat of the plate from lifting, and will save considerable work in flattening down afterward.

The mauls for this purpose should be of hard wood. Hornbeam is the best as it requires no banding as hickory or oak do. They should be 5" or 6" diameter, 10" or 12" long with hemi-

FLANGING.

spherical ends. The helves should be the same as are used in sledges.

When the turn is down, take a large faced flatter, and flat off the straight part of the flange. Use a small sledge for this or the flatter will make its mark. Flatters should be provided with iron handles as wooden ones are burnt very quickly. They should encompass the tool and not run through an eye. See figure 79. The faces of such flatters

Fig. 79.

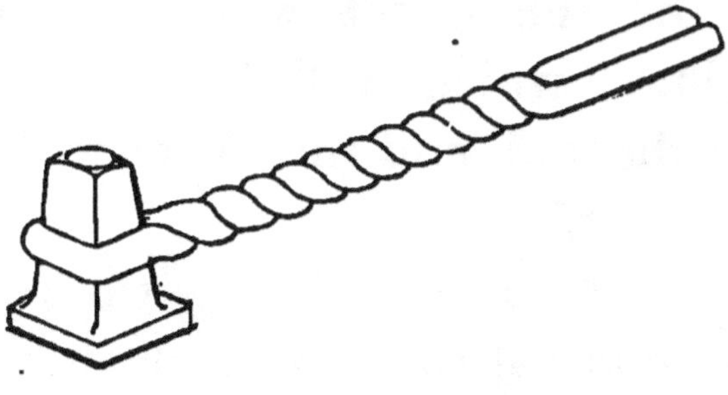

Fig. 80.

should be about 3" square, and have the

edge rounded off all around, commencing about ⅜" from the edge, and turning up about ⅛", as shown in section in figure 80.

If by this time the plate has not grown too cold, throw off the weights on the flat and if it buckles up, set it down with a flatter, while a helper holds a large sledge or block of iron against the flange. But if the plate does not show a redness, put it back on the fire and reheat. Never strike it while it is black, or the effect will be that of "piening," which will be relieved as soon as heated and the plate will be as bad as ever.

While one heat is being worked off, have a man at the forge cleaning and rebuilding the fire so that it will be ready the moment the previous heat is done. A good flange turner will never let the heat out of his plate, when once started, until it is ready for the annealing furnace or layer out.

In place of weights to prevent buckling while flanging, some use wooden staves, 2½" to 3" square, sprung in between the plate and the overhead timbers. The length of such staves being just the distance from the timbers to the top of the former, the thickness of the plate allows the necessary amount of spring required to give the pressure. It is hardly necessary to remark, that if the shop is high, and high it should be, that these staves would be impractical as they would be too long and unweildy.

To flange over an anvil, small block of iron or piece of timber, the plate must have a line drawn all around it just where the turn of the flange begins to leave the flat. As *a a* in figure 81.

Fig. 81.

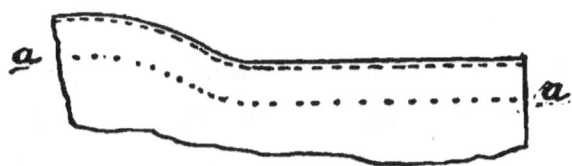

This line is a guide for a gauge, for which a large steel square may be used. A point being marked on its inner edges, each way, from the corner a distance equal to the outer radius. Figure 82 shows the method of marking and applying the gauge to the work. The mark, *c*, must follow the line of prick

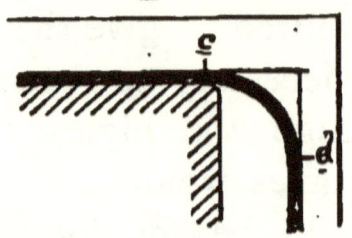

Fig. 82.

punch marks at *a a*, figure 81, while mark *d* indicates where the straight part of the flange begins.

It will be seen very clearly from figure 82 how much trouble and inconvenience will arise in working after this plan, yet a great many neat pieces of flanging have been done with it. The greatest care must be taken not to drive the flange too far, or it will necessitate turning over to drive it back.

Figure 83 is a throat plate of a locomotive boiler, and should be from 10 to

15 per cent., or the next sixteenth of an inch thicker than the rest of the shell, on account of the drawing of the metal in the concave flange. In an ordinary

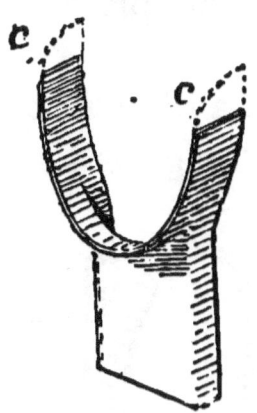

Fig. 83.

sized boiler with a flange 2″ inside radius and 4″ straight (for double riveted seam), the edge will be drawn in thickness as much as 1-16″. For the same reason, extra allowance must be made in the height of the rough plate, 5″ is usually sufficient. The corners, *c c*, having no resistance on one side, do not stretch as the middle portion does, and in consequence are pulled in. Before trimming, the wings will appear as in the dotted lines. This concave flange requires a special former. The side flanges may usually be turned on the former for the back head. Turn the concave first as there will be a better chance to hold the plate upon the former.

Figure 84 is a difficult piece of work,

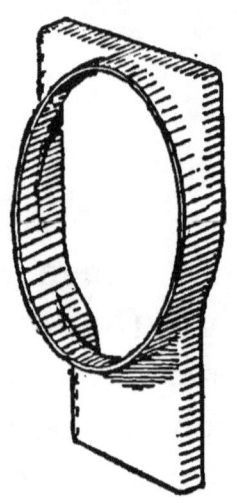

Fig. 84.

used to connect a "Belpair" fire-box to the barrel. Two formers were used for this; one for the hole, the other for the back side. The hole was flanged first. The former for the back had four lugs or brackets, cast upon its face,

FLANGING.

high enough to take full hold of the circular flange. These brackets served to steady the plate while the back flange was turned. Extra allowance of stock was necessary on the top and sides as those portions were drawn in by the flanging of the hole.

Domes should be riveted up before flanging except that portion of the lap that comes in the flange, which must not be punched until afterwards. Make extra allowance of stock on the corners for drawing. A line of points should be made on the inside, similar to those in figure 81, as a guide. The barrel plate should be rolled up to shape and used as a former to fit the dome by. After the dome flange is fitted, it can be used in turn as a former for the barrel plate.

VII.

WELDING PLATES.

To weld plates the greatest care should be taken to prepare the surfaces, as well as to get the proper heat. A perfect weld is flawless, the two parts being united equally throughout the length of contact, and they must be brought together before they will unite. All scale and cinder must be expelled. Scale, in the presence of a flux, and cinder, while hot, are fluid, and it is obvious that if the surfaces are concave or have concave spots, that when they are brought together, the sides of the cavities will unite and imprison them before there is a chance for them to escape, and no matter how hard the metal is hammered or pressed together a flaw is inevitable. Figure 85 shows a method

Fig. 85.

of preparing plates for welding given in an English work on boiling-making. It appears as if intended to catch all the cinder and scale possible. It is described as follows: "Thicken both edges, and split one and taper the other as shown in Fig.   , which is a section or edge view. When this is done, force one into the other, and close them over. * * * * Then secure the two plates by two or three pieces of angle iron bolted to them, also two stretching screws, and then place on the fire. When the plate is hot, the expansion of the metal acting against the stretching screws, will weld the plates even before they are hammered on the block, which of course must be done when sufficiently heated."

All of which is correct as far as it

goes, but if this splitting and tapering is done by hot chisel, hammer and swedge, there will be a greater or lesser number of little pockets to hold the cinder and scale. The very hammering down of the edge before placing on the fire will serve to confine the old scale formed in preparing the joint. Unless the joint were made a perfect fit by planing, a safe job could not be made; and even then cinder may enter because of the jaws of the split plate expanding a little before the tapered one expands to meet it.

About the only way to join plates with the ordinary tools at command of the boilermaker is to make a scarfed lap-weld as in figure 86.

Fig. 86.

In placing together allow them to shut by sufficiently to make a little thicker than the original plates. This to allow for the natural waste of the fire and to be certain not to go under thickness. The appearance of the weld before dressing down in finishing will be like figure 87.

Fig. 87.

It will be noticed, in figure 86, that the welding surfaces are convex. This allows the middle of the joint to make contact first, and by the gradual closing from the middle to the edges, it squeezes out all scale and cinder that is between. It must be born in mind that the edges of the joint must be the last part to be brought together.

It would be well before strapping the plates together to go over the welding surfaces with a coarse file or piece of gritstone to clean off all the old scale formed during the scarfing.

To weld the longitudinal seams of cylinders, the same kind of scarfed lap should be made. The plates being previously punched for the transverse riveted joint, to within ten or twelve inches of the scarf ; rolled up to shape and strapped inside with stout bars. The plate coming on the inside of the lap, to be rolled to the exact circle while the one on the outside must be left large after passing the punched section.

Figure 88 shows the form in which to get the ends of the plates, the thickness being exaggerated to show it more plainly.

Fig. 88.

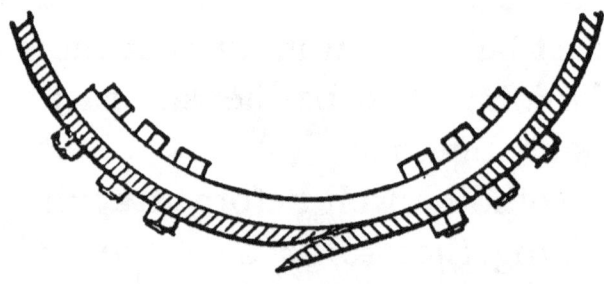

To prevent distortion when cold, it is best to heat the full length of the joint at once, but only hammer down a short section at a time, say eight or ten inches, and reheat. Commence in the middle and work both ways.

If the circle is large enough to admit of swinging the hammers on the inside, a block of iron, of a shape to fit the outside when finished, will be needed for an anvil. If too small for this, then a stout mandrel, or stake with face to conform to the inner surface of the ring, when finished must be used.

The writer has seen a cylinder 24" diameter, 36" long and 5-16" thick, with longitudinal joint welded in this manner. Tested under a hydrostatic pressure of 1,250 pounds per square inch, it did not show the first drop through the welded joint. It failed from the bursting of a head through a stay rod hole. After receiving a new head it was subjected to an air pressure of 850 pounds per square inch, and after three weeks it had leaked down but 7 pounds.

It is particularly necessary for welding to have a perfectly clean fire. No fresh coal, but simply coke should be used. The heating must be done gently, so as to give time for it to pass through the plate to the welding surface. Iron should come from the fire at a white heat; steel at a bright cherry.

The first blows upon the weld must be very light, and not till the plates are united should very heavy blows be given. As soon as the union takes place finish off as fast as possible with flatters.

VIII.

ANNEALING.

All plates, particularly of steel, upon which flanging or welding has been done, should be annealed. It is not so imperative for flanged iron plates, except such as are in contact with the fire, but all welded plates should be so treated.

To anneal, get the plate evenly heated to a glowing red. It must then be allowed to cool slowly, and as evenly as heated. Cover it with ashes or some material that will prevent cold currents of air from striking it. To get the heat a special furnace is the best and cheapest, if much work is done. It is specially economical where boilers of the locomotive type are made. Lacking a furnace, support the plate at its corners with brick, about a foot above the floor, and build a good even fire of. wood or charcoal, (charcoal is best) under it. The flange must be down. Keep a constant watch upon the plate, and if it reddens up in spots, quickly shove the fire aside that lies directly under them. And conversely if the plate reddens up generally, leaving dark spots, get hot coals under those spots as soon as can be.

Have a bed of dry sand made perfectly smooth and level, for the plate to cool off upon. Turn the plate, with flange up, upon this bed, and cover as quickly as possible with hot sand or ashes. Do not drop the sand on in shovel-full heaps, but distribute the sand, as thrown, over as great a surface as possible without stopping to spread it.

A good arrangement for covering a plate while cooling is a box made of No. 8 or No. 10 iron, inverted over it.

This will save a great deal of shoveling with just as good a result.

It being impractical to make a furnace arge enough to take cylinders, they must needs be heated by an open fire. Stand it on end, a little above the floor, and build a fire inside and out against the part that is welded. It will not be necessary to go all around it. Cool off by allowing the fire to die out, screening it from cold drafts of air. Be very careful to prevent an abrupt termination of the heat.

The general appearance of an annealing furnace is that of the one ordinarily used for reheating; the difference being in the size of door and bed as well as in the arrangement of the flues. Figures 89 and 90 are respectively a longitudinal section and cross section through the flues. The door being so long is made in two parts. One would be cumbersome and difficult to manage from its liability to warping from the heat. The flues are so arranged that

118 BOILER MAKING.

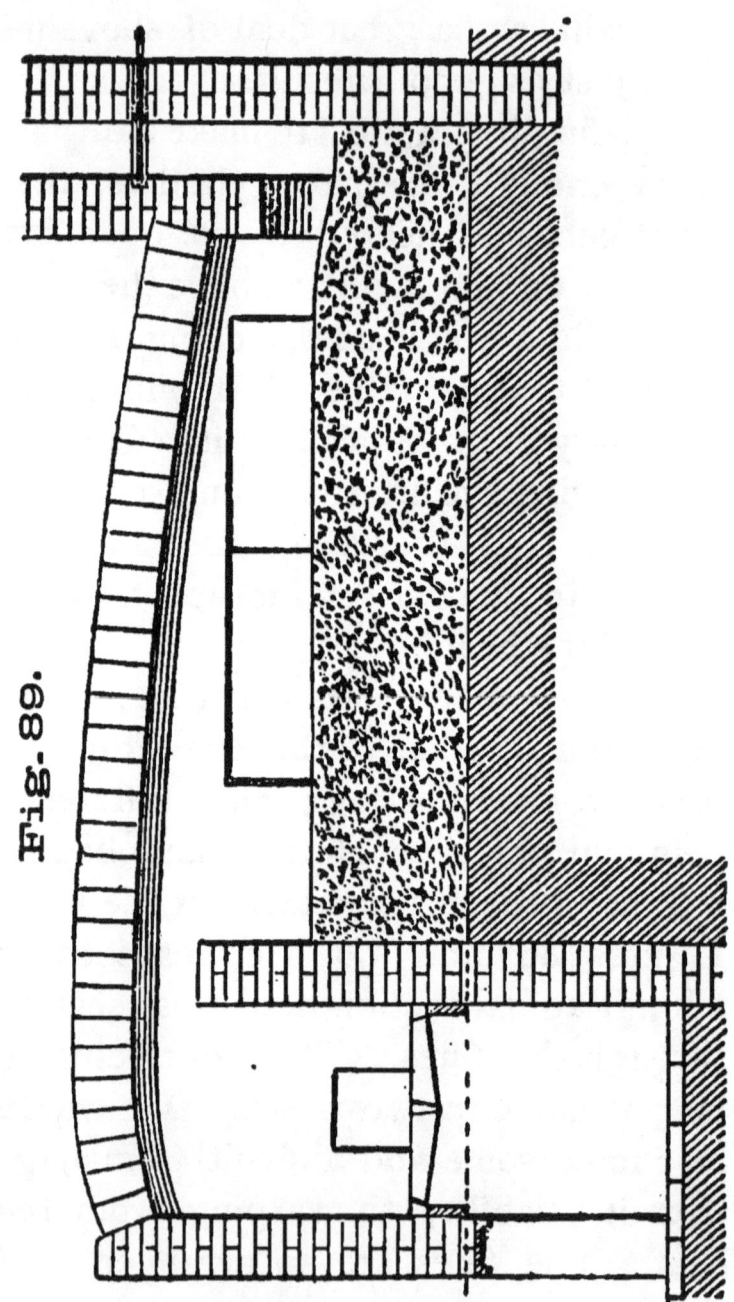

Fig. 89.

by the use of the dampers the play of the fire upon the plate can be readily controlled. A peep hole being provided in the door, one is enabled to watch the plate, and if one portion seems to be receiving more heat than

Fig. 90.

another, the damper in the flue opposite is closed, compelling the heat to go through another flue and thereby causing it to pass over the less heated portion of the plate.

In the furnace from which the cuts are taken the door opening is 5' 6" wide. The inside being 7' 0" by 8' 0. The grate is 8' long by 2' 9" wide. The bridge wall 2' 6" above the grate and 18" above the bed. From top of bridge to roof at centre 9", and at sides 3". The bed is of sand and 2' 3' above the floor. The flues are about 16½" wide and 10" high.

IX.
SMITHING.

It being impossible, in a work of this kind, to give a description of the methods of forging all the multifarious details that are used in boilers, a few of the principal ones only are given, as hints of what can be done by a little outlay to save a deal of hard work and a much greater expense, at the same time making a more uniform product, thereby saving a further expense in chipping and fitting.

Where two laps intersect, the corner of the plate that comes between must be drawn out to a taper to prevent an abrupt offsetting of its mate.

Never use a "dutchman" instead of tapering. It will be a trouble from the first—not only to keep it in place while riveting, but to get tight when calking. The plate should be all trimmed and punched except the corner to be thinned. The holes in this portion must not be punched before setting up, but drilled and reamed

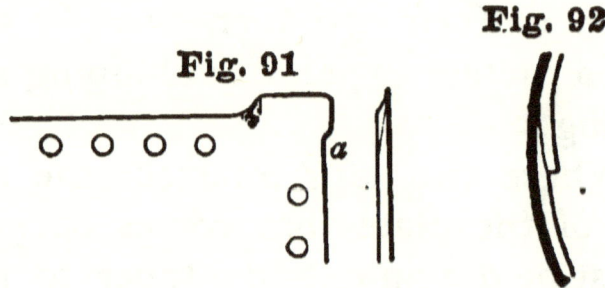

Fig. 91   Fig. 92

when in place. The tapering should take in the full width of lap and be drawn to as thin an edge as possible. The corner of the plate is shown in figure 91, and a section of the joint in figure 92.

Keep the plate level on the anvil, and after drawing, dress it down with a flatter. If it is very rough or of an uneven surface it will be almost impossible to prevent leakage. Allow it to spread a little on the edge *a*. This edge should not be trimmed until the plate is riveted into its place and ready for calking.

Curved crown bars are awkward and expensive forgings to work up without "formers;" especially so if

Fig. 93

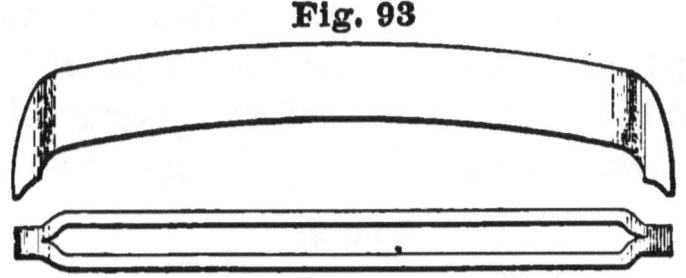

the ends are welded together. Take for example the bar shown in figure 93. It is well to have two formers for such, one for the ends and one for the general shape. It will be understood that this bar is formed in two parts and welded together at the ends after shaping them. The ends of the bars are trimmed roughly to the shape in figure 94, the point to be about $\frac{1}{2}''$ wider than the finished toe.

Have a heavy former made like

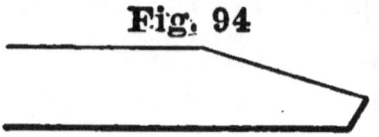
Fig. 94

figure 95; the opening being $\frac{1}{16}''$ wider than the thickness of the bar. The outside and the bottom of the opening must be just the shape of the end of the bar before giving the crown curve. The two lugs on the

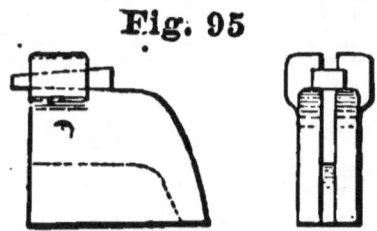
Fig. 95

top should be pretty heavy. The wedge must drive as shown in the figure. If driven in the opposite direction it is apt to jar out. The operation of this former is so obvious that further description is unnecessary, except to say that it is of rough cast-iron, needing no cleaning except in the slot.

The former for the complete shape is shown in figure 96.

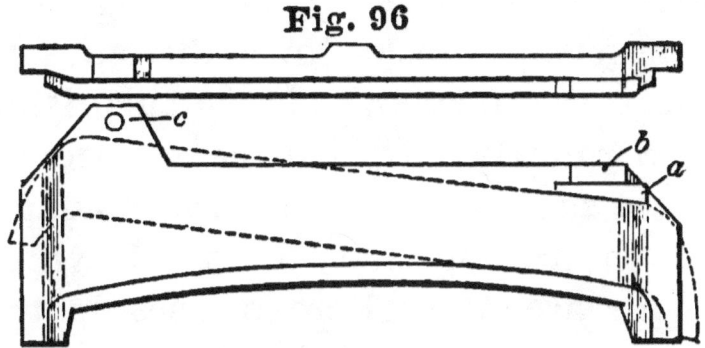

Fig. 96

Heat the bar throughout. Lay it on the former in the position shown in the dotted lines. Key it up tightly by the key *a*. Then with a cam-lever,

Fig. 97

shown in figure 97, which is pivoted by its pin in the hole *c* of the former, the bar can readily be set to its curve; the lever being swung first to the inside and then to the outside. The lip of the former must be made the same height as the thickness of the bar, to allow of flatting down in case of buckling. While the bar is being held to its curve by the wedge and cam-lever, knock down the ends to meet the former. It is surprising how quickly bars can be shaped by these contrivances. A furnace adapted to heating such bars, of which there are several good ones in the market, should be in the plant of every boiler shop. They are not only useful for this particular purpose but for a great many other details.

After getting the crown bars into shape, it requires but the welding of the ends to make complete. Do not attempt to weld both ends at the same time.

To make a "flanged" tee like figure 98, first double up the plate, setting the fold down close. Then drop the folded edge into the slot of a former like figure 99. Spread and flat dow the edges. The depth of the slot must be about $\frac{1}{16}''$ greater than the height of stem required. The stem will lose about that amount in cooling. The wings of the former must slope away from the slot $\frac{1}{8}''$ in $6''$. This is exaggerated somewhat in the cut to show it more plainly.

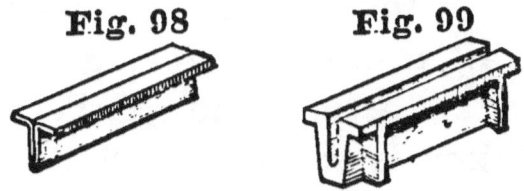

Fig. 98   Fig. 99

If curved tees are required, flange them first in the straight former, then bring them to the curve desired in a second former of the proper shape. Do not attempt to make them entirely with the curved former, as it is impossible to drive the stem down to the bottom for the full length; for, no matter how hot the metal is, it will persist in buckling. If the tee is intended for the waist of a locomotive boiler, where the rise of the wagon top causes the flanges to be twisted, it is best to fit them separately into their places. Having the centre of the tees made to the proper curve, very little work will be required to fit them, and it is impossible to make a boiler that will be exactly true to the curve called for. Indeed, some boiler makers fit the tees to the boiler from the straight, not attempting to curve them at all beforehand. Others think much time is saved, especially where the doubling, flanging, and curving are all done at one heat; and it can be. The inside of a boiler is an awkward place to work, and the less required there the better.

To make an eye or jaw upon a brace, two methods may be taken; viz., upsetting the rod until enough stock is made up, and welding on a stub of larger section, or smaller rods doubled around and lap-welding on. Ordinarily, welding in boiler work is objectionable, and should not be allowed where it can be avoided, unless upsetting is the alternative, especially in this matter of brace jaws and eyes. Much as has been said and written to the contrary notwithstanding, a properly welded eye or jaw is as good as, if not better than, the upset ones. By "properly welded" is meant not only that the welding should be well done, but that the joint should be properly designed. It may be asked, What designing can there be in a weld? The answer is, that there is just as much designing required in making a weld as there is in any other detail in the boiler. Take for instance a $1\frac{1}{8}''$ rod. Upset the ends to be united a little, and draw them out to the usual angle of 45 degrees, and we have a surface for welding of about $1\frac{1}{2}$ square inches area. Now suppose the joint was badly welded, and has a flaw in it of one half the area, which leaves $\frac{3}{4}$ square inch area, or about 25 per cent less than the cross-section of the rod. Now change the design, and instead of drawing out to the usual angle, the upset is drawn to a taper of 2″ in length, and the ends are lapped 4″ and welded. This gives an area on the joint of not less than $3\frac{1}{2}$ square inches. If then one half of the joint is a flaw, we still have more than double the area of the rod, unless the whole of it is at one end, which is very unlikely. Eyes and jaws made up on this plan of a long weld have proved themselves under test to be stronger than the rod.

On the other hand, to upset sufficiently to make stock enough, if the rod is at all fibrous (and what rolled metal is not?), the fibres are so driven back and buckled upon themselves that more or less of them part company, leaving numerous little flaws, and the fibres in any direction but that of the strain.

A short time since the author made

a comparative test of the strength of jaws made by the two methods. Having no proper testing machine or dynamometer, the only result attainable was where breakage occurred. The rig used was composed of a stout lever and a lot of pig-iron. Six jaws of each kind were tested. They were all practically of the same cross-section in the rod and around the eyes. In every case the made-up and welded jaw broke the rod, while but one of the upset ones broke there. One upset failed between the eye and the rod, at a point apparently the strongest. The other four failed at the eye, the failure in each case being so sudden that, although the closest attention was given to it, it was impossible to say whether the rent started from the hole or from the outside. On placing the pieces together they fitted perfectly, showing that no set whatever was given by the strain before or during the breaking. It may have been that the square rod used in the made-up jaw was of better material than the rods; but the price, which is usually a good indicator of quality, said it was poorer, it costing one-half cent per pound less. The rods were all of the same batch.

The jaw most in use is made of $\frac{5}{8}''$ square iron on a $1\frac{1}{8}''$ diameter of rod, with the eye $\frac{13}{16}''$ diameter. In fact it is used so often that it is practically

Fig. 100   Fig. 101

the standard in land boilers. See figure 56, *ante*.

To form this jaw take a $\frac{5}{8}''$ square rod about 30'' long, and double it up in the middle as in figure 100, and draw the ends out tapering on the flat way for a distance of 3'' or 4''. Double again in the transverse direction with a $\frac{13}{16}''$ pin in the fold. Let the ends remain open as shown in figure 101. Prepare the rod by upsetting a little at a point distant from the end equal to the length of taper given to the jaw. Draw the end to a wedge shape, leaving the point ¼" thick. It will appear then as in figure 102. Now bring both parts to a welding heat. Be careful not to allow the heat to run too far up the jaw part. Keep the pin in the eye as you weld up, so as to prevent its closing. Weld

Fig. 102 Fig. 103

with a rush, and do not tinker over it to make it look pretty. Every blow given after it loses its redness weakens it. Be very sure that the rod lines with the centre of the eye. It will now appear as in figure 103.

The next operation is to open the jaw; which can readily be done by heating close to the weld, and spreading enough to allow a block of a thickness to correspond with the foot used, and having a hole at the proper distance from the edge for the pin to pass through. Drive the block in so that the pin will enter and flat the jaw down.

In making a crowfoot brace, never weld the rod to the foot by "jumping" it on. It is very apt to leave unexpectedly. The jar of riveting will frequently start them. The most secure method is to hot-punch a hole through, and upsetting the rod, thread it through and weld up on the back side; the parts before welding appearing as in figure 104.

Fig. 104

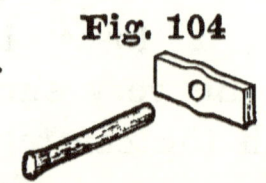

An anvil block with slot to allow the rod to drop into, so that the edge of the foot is below the top of it, also having a sufficient face to lay the foot on, will be very convenient for this work. Such a block is shown in figure 105.

Fig. 105

To weld the rod to the other foot—or hand, as it might better be called—allow it to lap about 3″ or 4″. Upset and taper the rod as shown in figure 102. The foot itself requires no preparation.

About the only place where welding by jumping is allowable about a boiler is at the corners of fire-box rings, or casing bars as they are sometimes called. As there is comparatively little strain upon them from the steam pressure, the great point is to get them water-tight. Jumping, if carefully done, will do this quite as well as if a more expensive lap-weld were made.

Cut the long bars to the outside length required, and jump short pieces upon the ends, so as to make two half frames as shown in Fig. 106. The short pieces on the ends must be long enough to allow for welding when the two halves are brought together. Trim and finish the corners

Fig. 106

completely before uniting the halves. By making up this way in halves much time is saved, as the work is lighter and of course easier to handle while finishing the corners.

In welding together allowance must be made for the shrinkage in cooling. This is an indefinite amount. A good average allowance is $\frac{1}{16}''$ per foot from side to side when the work is red hot. If on cooling it is found not just right, a very little drawing or upsetting will be required to bring it to the proper dimension.

In some locomotive boilers where the axles are under the fire-box, a recess or pocket is required on the outside of the box. This necessitates cutting the bar out for it. This can be done with hot-chisels, but it is as much as fifty per cent cheaper to have it done on a slotter. This element of cost must be reckoned not only from the smithing out, but from the fitting in of the plate and calking, the machined spot being square and curves fair. Indeed, some railroads require these rings to be machined all around inside and out, which makes a splendid piece of work, but rather an expensive one.

X.

PUNCHING.

PUNCHING, besides the necessary machines, punches, and dies, requires plenty of room and light. Convenient contrivances for handling the plates are indispensable for the quality as well as the quantity of the output. No satisfactory result can be obtained where it requires three to a half dozen or more men to hold the plate up to the punch by main strength. This is a method often seen in boiler shops. Nor is it any great economy to hang a plate up by chains or bailhooks and require one or two men to teeter it around while hunting the center point of the punch with a slate-pencil line on the plate, and that line half rubbed off; though this is a great improvement on the first-mentioned method.

A much better arrangement is to have a good-sized table surrounding the machine, with a ledge all around it $1\frac{1}{2}''$ high. On this table place a number of cast-iron balls about $2''$ in diameter; the height of the table to be such that the balls will support the plate to be punched just clear of the die. A plan showing about the relation of the table to the machine is given in Fig. 107.

The table should be plated with $\frac{1}{4}''$ iron to make a better surface for the balls to roll upon. The ledge around it should also be covered with sheet-iron to save wear. Between the inner edge of the table and the front of the machine there should be a lifter, operated by a treadle-lever, reaching to the front of the table, for the purpose of lifting the plate to meet the center point of the punch. It is a fact proved by experience that two men can handle a plate on this table that would require ten or a dozen men to place there, and punch from twenty-five to

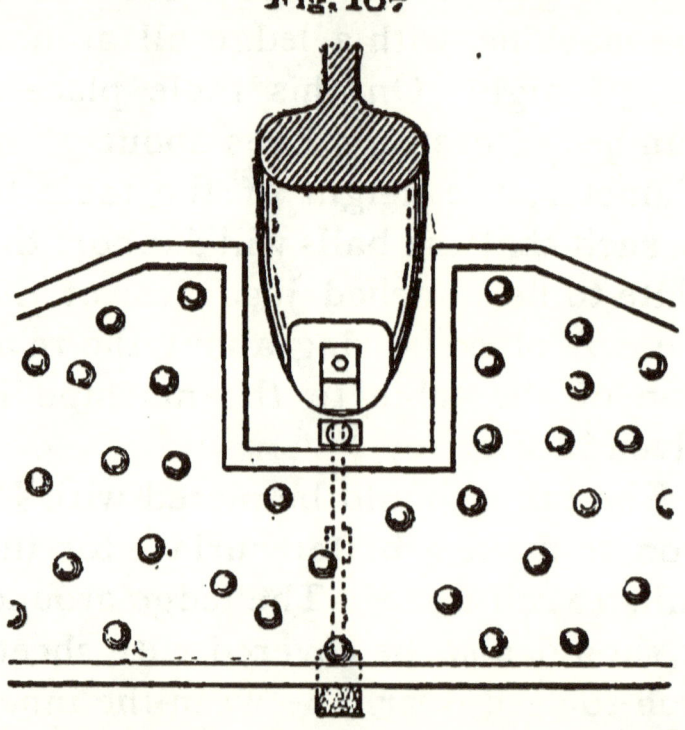

Fig. 107

fifty per cent more holes than by the aid of any other arrangement. This table can be built at an outside cost of thirty-five dollars.

In preparing a plate for punching,

the center lines for the holes should be very distinctly marked, and every center gone over with a prick-punch. Give the prick-punch a light trial tap at first to see that the proper place is hit. If it is, give the punch a good solid blow, taking care that the dent is not made so large that the centering point on the machine punch will float around in it. If the proper spot is not hit at the trial tap, cant the prick-punch over and drive it sideways until the point reaches the proper place and then holding it against the shoulder at the end of the little groove cut by the drawing over, give the finishing blow.

If great care is not taken in centering the holes, it cannot be expected that the punching will be first class. The machine hands have quite a plenty to do to steer the plate to the center-punch marks without being called upon to climb all over the plate to see whether the punch will come, down in its proper place when the prick-mark is out.

In case there are a number of plates of the same size, it would be well to have the rows of rivet holes laid off on wooden templets $3''$ or $4''$ wide and $\frac{1}{2}''$ to $\frac{7}{8}''$ thick, and have the centers bored $\frac{3}{8}''$ diameter. These templets can be clamped upon the plates to be punched, and a prick-punch, turned to fit the holes, can be used with great rapidity. These templets should be saved as records, or to use in duplicating a plate in case of repairs. For tube plates a board templet covering the whole plate so as to be set by the edge of it, and having center holes to locate all the rivets, as well as tube holes, is a desideratum even for a single boiler, as by its use on both plates it will be almost impossible to get the tubes in crooked.

A bad practice in many shops, and one that cannot be too heartily condemned, is the marking off of rivet holes with a bit of pipe dipped in white lead. Proper enough, certainly, where the punch has no center point to center by, but it should be used in no other case. It is a good plan, however, to go over the plate with it after the prick-punch has been used, marking a circle about each center, as it will save time on the machine, more especially if the machine is in a dark corner or the day is dull.

The plate should not be trimmed before punching, as it frequently happens that some few holes come just on the edge of the plate, and the punch is in danger of breaking if an attempt is made to punch half a hole. It is very easy to trim off to the center of the holes after the punching is done.

In punching heavy plate and light plate with large punches, it is necessary to keep the punches cool with oil or oil-soap, as the heat caused by the compression of the metal is very apt to draw the temper of the punch.

The faces of punches for plain holes are made flat, flat with a centering point or having a spiral shearing edge. The two first named are alike, but for one having a centering point. The flat face without the centering point is fast becoming a thing of the past. More time is wasted in setting a plate for a hundred holes than would pay for the punch with the centering point. Except for the centering point the faces of both should be perfectly flat. If made hollow, the edge is weak and is liable to break off at any time; and if made convex or conical, it requires much greater power to drive it through the plate, and a larger opening in the die as well.

The spiral shearing punch, Kennedy's patent, has a face cut out in such a manner as to form a spiral on the edge, as shown in figure 108.

The figure shows a double cut, such as is preferred by the patentee. There have been punches in the market with but a single cut, but they are apparently infringements on Kennedy's patent. The peculiar advantages of the spiral face is that it acts in the manner of a shear. The only portion of the face that actually punches is the narrow flat portion which backs up the entering edge. As the punch shears or cuts the plate instead of

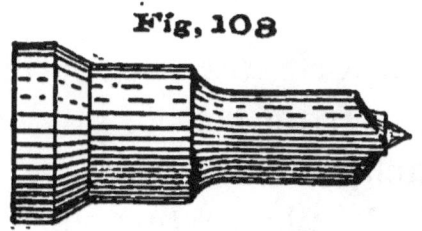

Fig. 108

rending the fibers, less strain is thrown upon the machine, and the plate about the holes is strained to so slight a degree that it is unnecessary to anneal. From experiments made by the chief surveyor of Lloyd's Register of British and Foreign Shipping it was shown that the ductility of plate punched with the spiral-faced punch was over 50 per cent. greater than when punched with a flat-faced one, and the power required to drive the punches was nearly 50 per cent. in favor of the spiral-faced. Also in several experiments where two holes were punched in the same test piece, one with the spiral and one with the flat, the test piece broke invariably through the hole punched with the flat punch.

The author's own experience has shown that from lack of power a $1\frac{5}{16}''$ flat punch could not be driven through a $\frac{7}{8}''$ plate; but when a single cut spiral punch was used, no difficulty whatever was found. Holes as large as $5''$ diameter have been made with a spiral punch, and with no more apparent strain on the machine or greater power used than shearing a straight cut of a length equaling its circumference.

Punches should not be made on a stock, but should be cut short and united to the stock by a coupling nut.

This will save great trouble in changing punches, as the stock and nut may be made to suit a large punch, while the butts of the smaller ones can be made large enough to fit the same. Figure 109 shows the arrangement with nut in section.

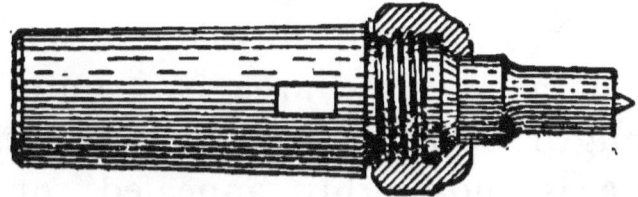
Fig. 109

Punches should be made tapering $\frac{1}{32}''$ per $1''$ of length for clearance. Dies $\frac{1}{16}''$ per $1''$ on the same account.

The diameter of dies for the minimum power to drive the punch, as determined by Messrs. Wm. Sellers & Co., may be found by the formula

$$D = d\,(.2t), \quad \ldots \quad .8$$

where $D =$ diameter of hole in the die,
$d =$ diameter of punch,
$t =$ thickness of plate.

This makes a considerable taper in the hole, which is objectionable because of the loss to the plate left standing, with no gain to the rivet's strength. It also requires quite an amount of upsetting to make the rivet fill the hole. Some boiler makers use a die of but half the clearance given by the formula, under the mistaken idea that the gain in plate area is a gain in strength: which is not so, unless the plate is thoroughly annealed, or a spiral punch is used. The patentee of the spiral punch uses the formula

$$D = d + (.1t), \quad \ldots \quad 9$$

finding so little difference in power required, that the gain in plate strength was equal to the amount saved.

Tube holes should not in any case be punched to full size. With a spiral punch allow $\frac{1}{16}''$ in diameter for riming, and with a flat punch have the hole no larger than is necessary for the teat on a counterboring tool.

PUNCHING. 149

For the riming and counterboring of tube holes see Chapter XI.

All punched holes should be laid off on the faying surface so that the holes may be punched from that side.

Fig. 110

They should also be countersunk slightly on the outside to take off the wire edge left by the die. Figure 110 is a section showing a properly fitted hole.

XI.
DRILLING, ETC.

In boiler making, as well as in machine work, there is no one operation upon which the condition of the tools has so great an influence as in drilling, and its kindred operations counterboring, countersinking, and riming. The tools should not only be sharp, but of proper form. The angles and equality of the cutting edges should be particularly looked after.

For drilling, the "twisted" tool is the best and cheapest—not for first cost, but for the amount of work obtainable. Once made, it can be worn, ground, reworn and reground, until nothing but the stock is left, and it is always to size and the cutting

DRILLING, ETC. 151

edge just the right angle for the best effect; whereas the flat drill is a constant source of expense for dressing, to keep to width, and is very difficult to grind to the proper angle, seldom being much better than a scraper. To be sure it requires a good mechanic to keep twist drills in order, but the difference in the result obtained will more than pay for his

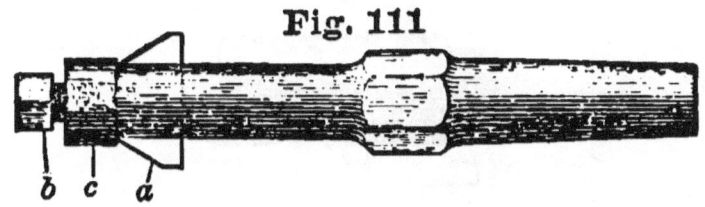

Fig. 111

keeping. It would be well to have all small tools kept in order by one competent man, allowing no other person to sharpen or dress them.

Countersinking tools should be made with a teat, so that they will always work central. They can be made with a stock to fit the same socket as the twist drills. A neat and effective tool is shown in figure 111, which is a copy from one designed for $1\frac{13}{16}''$ holes, and was very handy and economical. The cutter a is $\frac{1}{8}''$ thick, $1\frac{1}{2}''$ long, and is kept in place by the set-screw b, which is $\frac{3}{8}''$ diameter. The collar c is driven on, and if worn can readily be renewed. The two parts cutter and collar are the only ones subjected to wear, and for

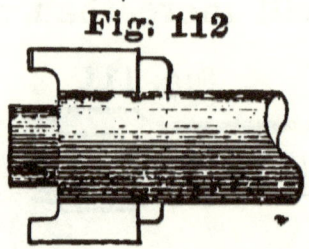

Fig. 112

the amount of work done by them the cost of maintenance is merely nominal. An important point in this tool is to have the stem a little smaller than the collar, so that the chips may have a clearance. In the one spoken of the body was $1\frac{1}{16}''$ diameter.

For counterboring tube holes a tool like figure 112 is used. This also has a collar on the point to allow of renewal for wear. The cutter is notched out to fit the stem, like a gib, so that it may center itself. The slot in the stem must be long enough for the cutter to pass through easily. The stem should be pretty stout, as the cutter to be stiff enough to stand up to its work must be not less than

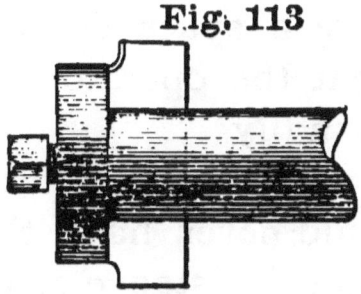

Fig. 113

$\frac{1}{8}''$ thick for $2''$ holes or less, and $\frac{1}{16}''$ thicker for every inch larger than $2''$.

For rounding the edges of tube holes the same style of tool may be used, with the cutter made in the proper shape; the cutting edge starting flush with the side of the collar, as shown in figure 113.

The drilling machine should be an overhead one, leaving the space all around clear for swinging the plate.

The feed is best worked by a rack and pinion movement, and operate, through a levis wheel on the pinion shaft, by a continuous chain with hooked shifting weight.

A ball table such as described for the punching machine is a good thing to have under the plate drill. It should have a space cut out on one side, so that the operator may get to the drill handily.

Fluted rimers for holes in plate metal should never have the cutting edges an equal distance apart. Were they equal, it would be almost impossible to obtain a round hole. No matter how many flutes there are, if equally spaced they are sure to jump, so that a tap could not follow and make a perfect thread. No particular care need be taken in laying them out. A good way is to space them by "thumb" measure, and no trouble will be found in getting a truly circular hole.

XII.
TRIMMING.

ALL trimming of edges, except thinned corners, should be done before setting up. To chip edges with the plate in place, without injuring the under plate, is a slow and expensive process, and where it is impossible to avoid it none but the most careful and experienced hand should be trusted with it.

The angle to which the edges of plates should be trimmed is about 120 degrees. This is found from long experience to be the most convenient for calking, and chipping while in place. See figure 114.

All holes that can be, should be punched or drilled before trimming.

Never use a hammer and chisel when it is at all practicable to machine the edges of plates. For planing straight

Fig. 114

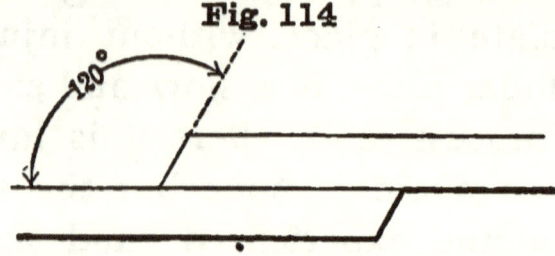

work the best machine is that on which the plate can be clamped down solidly in its full length, and the tool traversed back and forth by a screw. It is important that the plate should lie flat for planing, as the tool cuts the edge to an angle; and if the plate

Fig. 115

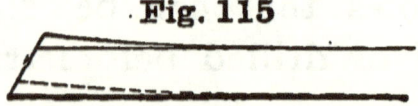

buckles up, it will cut deeper and thereby make the lap narrower than it should be. Or, if this point is taken care of, that part of the plate which is held down will be wider than is necessary. Figure 115 shows how this effect is produced.

Circular heads can be trimmed to advantage on an ordinary boring mill or large facing lathe; while the straight part of other flanged work can be trimmed on a planer. Too many boiler shops lack these machine tools. Where the boiler shop is connected with a machine shop it is far cheaper to carry a plate to the boring mill or planer, and have all possible work done by them, instead of chipping them. Even where the planer must necessarily leave considerable for chipping, it gives the chipper a better guide for width and angle, and neatness as well, for without thinking of it he will endeavor to emulate the work produced by the machine.

Corners where spread for intersecting laps should not be trimmed until after the plate is riveted into its place, as they frequently require a little upsetting to make them fill the space, and a "dutchman" should be an unheard-of makeshift in boiler work. Or it may be that a little more thinning is needed, in which case the work of trimming them previous to fitting in would be thrown away.

Flanged plates should be laid upon stout trestles or blocks, and have a few hundredweight piled on them to keep them steady. The trestles or blocking ought to be high enough to allow the men to stand to their work. Standing they can get a good free swing of the body to counterbalance the throw of the arm and hammer. A sitting posture, which is compelled by having the work too near the floor, is conducive to fatigue, as it throws too much work on the muscles about the waist, and cramps the legs so that a frequent letting up is required for resting and stretching. Keep the men on their feet, and they will do double the work and not notice it.

XIII.

COLD BENDING.

A PLATE should never be bent cold with hammers or mauls. The best of metal will be ruined thereby for use in boilers, unless it is afterwards annealed, and even then nothing but the very best can be relied upon. If the metal is cold short it is more than liable to crack through, and a hammer will cut such dents as to materially weaken the most ductile. The better way, if rolls are not at hand to do the bending, is to leave the boiler unbuilt, as in that shape it is certain not to burst, whereas with cold hammer-bent plates it is almost certain to burst unexpectedly, sooner or later.

Cold bending should be closely looked after, as the symmetry of the work as well as the fitting together of the rings of circular shells depends greatly upon the accuracy with which the bending is done.

The rolls should be stiff enough not to spring to any appreciable extent while doing their work. The upper roll must be arranged so that it may be easily removed by shipping it on end. To do this the upper part of the housing or frame at one end should be removable. A good way, that saves considerable work, is to have it so pivoted on one side that by slacking a nut on the opposite it may be swung sufficiently to one side to allow the roll to pass by.

Large cast-iron rolls should be cast with a central core to take out somewhat of the internal shrinkage. A solid roll will break with less than half the strain that it takes to break one having its center cored out.

The ends of the shifting roll should be fitted with eye-bolts to hook the lifting tackle to on swinging it in and out of place.

The rolls should have a reversible motion, so that the plate can be worked back and forth until it comes to the proper shape.

Having the plate punched and trimmed as far as practicable, it is then ready for the rolls. Raise the top roll sufficiently to allow the plate to run through easily. Carry it on to the middle, then screw down as hard as possible on the roll bearings, and before starting the machine see that the plate is set square to the rolls. A large wooden square of a parallel thickness, pushed into the angle formed by the top roll and the plate as a base, is handy to correct with. Cut a sheet-iron templet to the exact curve the inside of the plate is to be, and in length about equal to the radius. Roll the plate back and forth, screwing down on the roll bearing each time it is rolling in the same direction it started in—carefully testing it from time to time with the templet.

Take good care not to roll it small, for it is a difficult job to spread it,

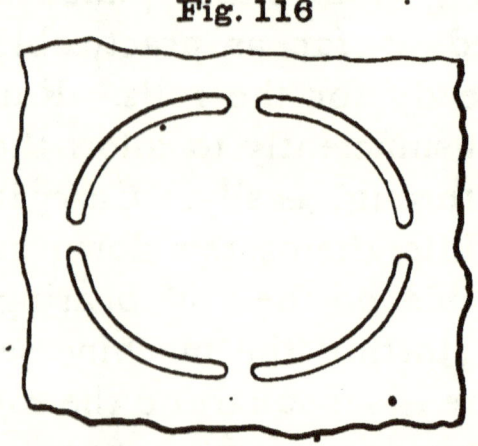

Fig. 116

more especially if it should be but a short section that is quick of the curve. If it should be left a little large, it can be sprung together by passing a chain loosely about it and drawing up the slack by twisting it with a bar.

Where large openings come in the

plate, such as for domes or manholes, the waste piece should not be cut out entirely before the bending is done. It should be left united to the plate by four short sections to be chipped out after bending, as shown in figure 116. If this piece is left out the plate will bend very short opposite the opening, as it will lose in stiffness directly in proportion to the space cut out.

Be very careful, as the end of the plate approaches the roll, not to let it run by, or it will snap out and fall so quickly as to endanger the lives of any within its reach.

XIV.

SETTING UP.

THE setting-up floor should be spacious, and provided with powerful cranes, chain slings, hand hoisting machines or tackle, portable stages, and plenty of blocking. Rivet-heating forges and one or two open forges should be disposed in convenient places. The roof framing should be strong enough to support a considerable weight depending from a fall or other hand hoist, unless there is a traveling crane which commands the whole floor. Over the power riveter there should be quite a tower, in the top of which a traveling crane of small range, say of ten feet or so, each way, for ordinary work. The lift should be considerable—not less than thirty feet. This crane should be arranged to be operated entirely from the floor, so that the operator may be near the work and see just how much or little to shift it himself.

In setting up a boiler it is important that every part is fairly matched to its neighbors, and also that it is properly in line. Nothing looks worse than a long cylinder boiler "swaybacked," like a country horse; or "hogged," because it was put together in a harum-scarum fashion; or a dome hanging to one side; like a gambler's hat.

The horizontal seams of cylindrical boilers should be riveted before the sections are brought together. On bringing them together it is frequently found that instead of being perfectly circular, they are somewhat oval, and do not match. To get over this, pass a long bolt through two opposite rivet holes, across the long diameter, and draw on them so that one section will enter the other. After sliding together till the bolt ends foul, take out the bolts, and if the sections still bind so as to be difficult to get them further, take a chain fitted with a hook at each end—bail hooks are the best—and hook over the outer edges of the two sections, and twist up the slack with a piece of scantling or wooden handspike. An iron bar is not so good, as the chain is apt to slip on it. A special contrivance for this purpose is shown in figure 117. An ordinary chain sling may be used in

Fig. 117

connection with the screw jack shown, though one with a "bail hook" is preferable to a plain-hooked one, as being less likely to mar the trimmed edge of the plate.

After getting the sections together

and lightly bolting through the rivet holes, the end sections should be leveled and lined with each other. Through a rivet hole in the top and bottom centers a plumb-line can be passed and the sections set vertically thereby. Having the end sections blocked firmly in position, it will be very easy to straighten up the intervening ones with a lining cord, starting with the next ones adjoining the ends. When two adjoining sections are properly lined, look for a few rivet holes that come exactly fair. If the laying out and punching has been done in a reasonably proper manner it will not be difficult to find sufficient for the purpose. Into these holes drive tightly fitting bolts, and set the nuts up solidly. By these means it is pretty certain that the boiler will be straight, though it would be well to prove it by going over it again with plumb-line and cord. If all is straight, set in the heads, being careful that they are square with the barrel, and also that the lines of tube holes—if it be a tubular boiler—are level. Everything being in line, plumb, level, and square, put a clamping bolt into every third or fourth rivet hole, and clamp up tightly. These clamping bolts, and the steadying bolts as well, must remain until it is necessary to remove them to get in the rivets.

Where a machine is used for riveting, more than two sections cannot usually be put together at a time, as there is a limit to the length of the stake or "holder-on" of the machine beyond which it cannot reach. In such a case each section must be lined independently as it is put in connection with the next preceding one. This can best be done with a straight-edge, taking care that the section being lined in is not oval at its outer end. Draw in its long diameter, and hold it there until it is fully riveted in, if such should be the case. The head having its flange turned in should be first riveted in on starting. Boilers having one head with flange turned in and the other with flange turned out can have every rivet driven by the machine except, certainly, such as hold the stays to the shell.

In a locomotive type of boiler the fire-box must be completely riveted, the crown bars fitted and bolted on, and the calking all done on the outer or water side—the inner or fire side to be left without calking till all riveting and stay-bolt setting is done. The fire-box shell must be riveted up, except the front or throat plate. Then set them together. Tie them there with a few bolts running through the stay-bolt holes. Now see that the door flanges match. If the former was properly made, the male flange will be found a little small. To spread it to fit, have a number of iron blocks heated white, and lay them against the door flange. This will heat the flange in a very short time to a red, when it can readily be spread to fit. Do not attempt to spread if it does not show at least a red glow. Especially with steel a black heat is a dangerous one, and sometimes even a light blow will crack it. It would be easier on the plate to pien it out cold than to attempt to draw it at a black heat.

Before riveting the fire-box into its shell, have the barrel fitted and lined to it, and securely clamp-bolted. If the flat sides of the shell bulge outward, it will be necessary to draw them in with long bolts through the stay-bolt holes. If one side is flat or hollow while the other bulges, use a stiff piece of timber on that side, reaching from end to end, through which run the clamping bolts. The hollow side can be drawn out by the same piece of timber, by using short bolts having a bent end, instead of a head to hook through the plate. All such clamping arrangements must be kept on until all stay-bolts are set that can be without their removal.

The fire-box with its shell being set up, level fore and aft, and its center plumb, bring the barrel portion into place. If the boiler is of the raised wagon-top variety, the upper connecting or offset plate is not put in place until all the rest is properly set and firmly secured, when it can be tried in and the rivet holes marked on it. It is best to mark them with a centering punch, which is made to fill the hole in the other plates. It is then to be taken out, punched, and trimmed, when it can be put back permanently.

To get such a boiler in line sidewise place a straight-edge on each side of the fire-box, to extend as far as the forward end of the barrel; drop a plumb-line through the vertically central rivet holes, and measure from the line to each straight-edge. Longitudinally, level each section, or if the blocking is so arranged to get through pass a straight-edge under the fire-box and measure up from it the proper distances to the under side of each barrel section.

To plumb the fire-box, when the upper part is wider than the lower, throw a line over it with a "bob" on each end, and see that the distances from the line to the sides are equal.

Leave the dome off until all else is riveted, as it is very convenient to turn the boiler on its back while riveting and calking the under side.

Rolling over a boiler must be intrusted to none but the most careful and reliable men. A little carelessness may necessitate quite costly repairs. The author has seen a large steel locomotive boiler so badly injured by falling against another that the shell plate of the fire-box was knocked in, the stay-bolts driven through the inside plate, and the crown plate all buckled up. It cost nearly half the selling price to repair it. The riveting had all been done, and it was necessary to cut out the fire-box to get at the crown plate.

A fall or set of falls should be used on *both* sides in turning over—one for hoisting and the other for lowering. The lowering must be done slowly. If there is the least doubt of the strength of the tackle or the framing from which it is suspended, keep blocking under the boiler, taking out a little at a time as it descends, or adding as it ascends; allowing but a few inches fall in case of accident. Take no chances, for not only the boiler is in danger of injury, but life and limb as well,

Give the boiler no chance to slide sidewise on the floor as it rolls down. This is a fruitful cause of accident; for when it does happen, it is as sudden and vicious as the kick of a mule, and will try the strength of everything supporting it very severely.

In setting up a cylindrical boiler with furnace flues and back connection, like that shown in section in figure 16, page 34, have the shell and front head set up and riveted as for a plain boiler. The flue and back connection, set and riveted together, are then put in place, and the flues riveted to the front head. Be sure to set the back connection properly, so that the tube holes will line perfectly with those in the front head, otherwise there will be trouble in setting the tubes. Lastly, put in the back head.

XV.
CALKING.

THE object of calking being simply and solely to produce a "stop water," when that is attained no further benefit can accrue from it. But great evil may arise from over driving, to which there is a great tendency as usually practised. Besides its bad effect upon the work, it is a useless expense of both time and the hardest kind of labor.

There are two prominent styles of calking in vogue—the square and the concave. Figures 118 and 119, page 177, show sections of joints illustrating them, and the tools by which they are produced. Figure 120 is an ornamental style said to combine the advantages of both. Figure 121 is the same as figure 118, but has the acute corner slightly rounded, so that it will not cut the under plate. Figure 118 is the easiest to work, and the tool is of the least trouble to keep in order, but it will nick or groove the under plate with its sharp edge. The sharp corner of the tool is first used outward to "split" the plate. Then the tool is turned over and the work finished in the shape shown.

Figure 119 will not cut the under plate, but there its advantage stops, for it is very hard to drive; in fact, with heavy plate it requires several sizes—a small one to start the groove, and one of proper size to finish, with several intervening sizes to bring it up gradually.

Figure 120 appears to be as hard to drive as a smaller size of 119, while there is yet a square corner to mar the under plate. It is claimed that this tool makes very handsome-looking work.

CALKING. 177

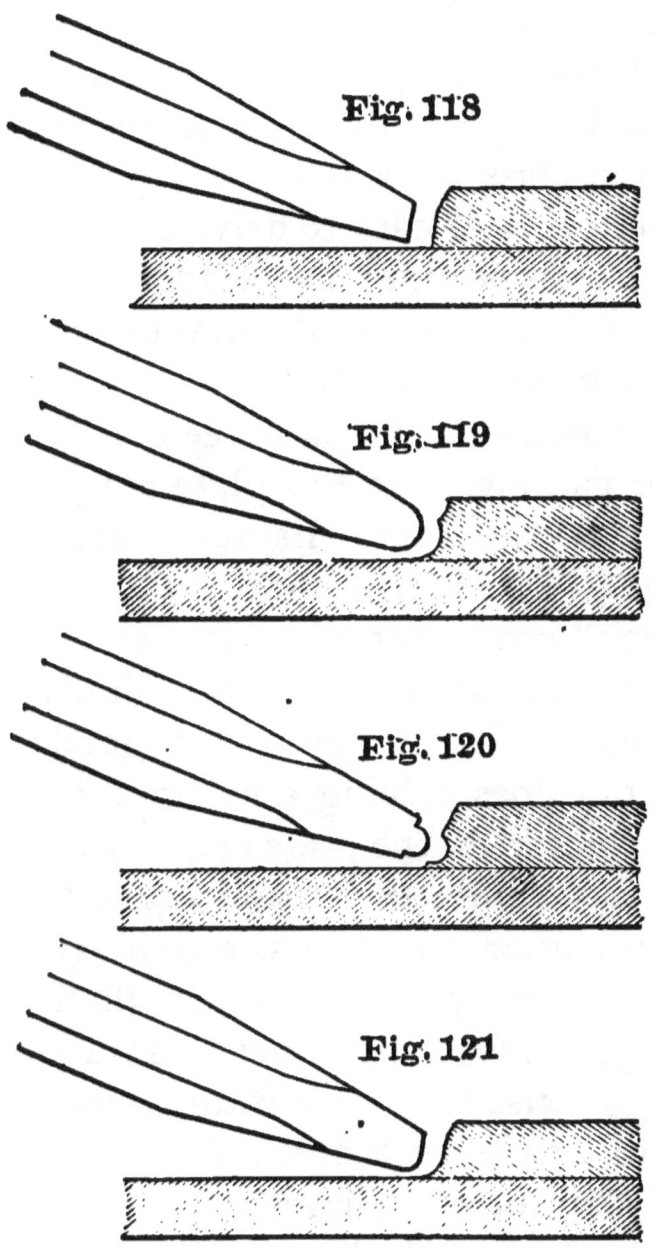

Fig. 118

Fig. 119

Fig. 120

Fig. 121

What good beauty does when covered up by brickwork or non-conducting material is a question; and covered up all first-class work is.

Figure 121 is as easy to drive as 118, and as it does not mar the under plate any more than 119, its advantages are certainly greater than that.

It is claimed by the patentee of the concave calking that the plate is upset to a considerable distance from the edge; "one half to three quarters of an inch" producing a bearing which "increases the strength of the joint." If this were a fact it would be a good thing; but does driving a wedge into such a place keep the plates together? Will it not rather throw an uncalled-for strain upon the rivets, which are already loaded pretty well by their shrinkage strains? But does this state of things exist? The author determined to let the work speak for itself. A single lap joint was riveted up from $\frac{3}{8}''$ steel plate, and three styles (figures 118, 119, and 121) of calking done upon it. It was then cut crosswise of the joint into six pieces; the faces left by the cutting were brought to a high finish, and treated with acid. No indication of compression whatever was shown. Believing this was because the steel was so homogeneous in structure that it would show to better advantage in iron, another test piece of iron plate was treated in like manner. The result proved to be entirely different to that claimed. The upsetting was in no case over $\frac{1}{32}''$ in depth, and in one case—opposite a rivet at that—there was quite a distinct opening of the faying surfaces with 119, which commenced opposite the crown of the curve, and reached a point opposite the corner of the rivet head. This opening may have been there before the calking was done, but it certainly was not closed by it, as it would have been had the upsetting theory been correct.

The compression of the metal in all the specimens was practically equal in depth, but with 119 the metal flowed or spread on either side, producing a long feather edge against the under plate, and a slight concavity on the high edge of the outer, as shown in figure 122.

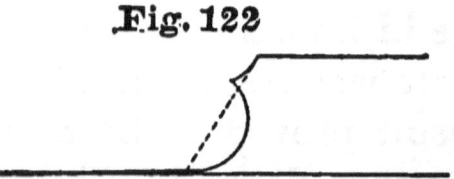

Fig. 122

The tools should be stiff and stubby, as short as is convenient to hold. Drive lightly, remembering that all that is needed is but to check the little pin-hole leaks, and that the rivets, not the calking, hold the plates together. To drive as some calking is driven is equivalent to driving an ounce tack into a segar-box with a sixteen-pound sledge.

XVI.
TUBE SETTING.

BECAUSE of the irregularities of the tube plates, which are inevitable in the best of boilers, it is necessary to measure out to out of them, for every tube—of course adding the extra length for beading. This measure is best taken with a piece of gaspipe, as it is so much stiffer for its weight than a solid rod or wooden staff, as well for its convenience for catching with a small rod at the farther end.

The measuring pipe should be held flush by the helper at the far end, while being marked close against the plate at the near. The helper will find that a block of wood held against the pipe while bearing on the plate will be much nicer than the thumb or palm of the hand. Use a slate-pencil for marking, for very little care is required to preserve it as long as is necessary, and file marks coming so near together will engender confusion.

Try all the holes first to get as many as possible before marking off any tubes. It will be found generally that the lengths run in clusters, that do not vary much over $\frac{1}{16}''$ among themselves, which is near enough to work to. These clusters can be surrounded with a chalk mark so that their limits can readily be found on putting in the tubes.

The length required for beading is about two and one half times the thickness of the tube, and should never be less than twice.

Where copper ferrules are required at the fire-box end, and the space between the tubes is less than one third their diameter, the tubes must be swedged sufficiently to allow room for the ferrule of same outside diameter as the tube. This swedging can best be done under a drop-hammer, though there are hand-power machines that do the work fairly well. The ferrules should not project much farther beyond the plate than just sufficient to give them a hold, and allow the tube to be beaded over them.

In placing the tubes, commence with the bottom rows and follow right along, the helper at the far end using a short rod for catching and directing his end into place.

The process of expanding should be a progressive one. Do not finish setting one tube before commencing another. Rather take one step at a time throughout.

First get the tubes fixed in their proper position. This is done by battering out with a straight-piened hammer—the pien being crooked downward; or, better, with a tapered triangular plug, the angles being rounded off to about the shape of the

Fig. 123

inside of the tube. Figure 123 shows such a tool. This plug makes a good expanding tool in default of a better one. It is slow work, however, as it requires so many turnings.

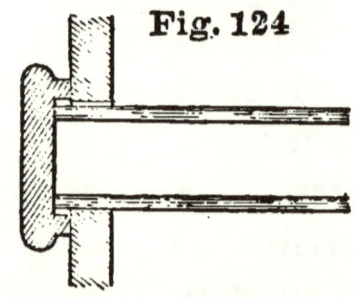

Fig. 124

While setting out one end tne helper holds the far end at its proper distance from the plate. For this purpose a small block of iron, with a recess in it of the proper depth, will be found very convenient. Figure 124 shows how it is to be used. The bead around its edge is for convenience in handling.

There are several good tube expanders in the market, the principal ones being the "Prosser" and the "Dudgeon."

The Prosser expander is made up of a number of segments held in place by an encompassing spring. The outside of these segments is shaped to such forms as it is desired to have the tube set to. The insides of them are made to fit a tapered mandrel, which is driven in with a hammer. This tool requires frequent loosening and shifting around on account of the spaces left between the segments.

The Dudgeon expander does first-class work in a first-class manner, being a series of small rolls, of proper shape for forming the tube, forced out by a tapered mandrel, which is kept turning while being driven in—thus spreading the tube with the least possible chance of splitting it.

The triangular plug shown in figure 123 sets the tube more on the outer side of the plate than the inner, and in consequence does not make as good work as if it were otherwise; yet the tube will hold the plate against the outward pressure of the steam as well as the best.

Fig. 125

Beading is done, after the tubes are expanded, by means of the "boot" tool shown in figure 125. The long leg goes inside of the tube. The beading must be done gradually, working down a little, all around, at a time, to avoid splitting. The tool shown should be cocked at first so that the long leg lies with its full length against the inside of the tube, and gradually as the end spreads it is

TUBE SETTING. 187

brought out so that the "heel" will set the edge close against the plate.

Where the tubes are large, and the

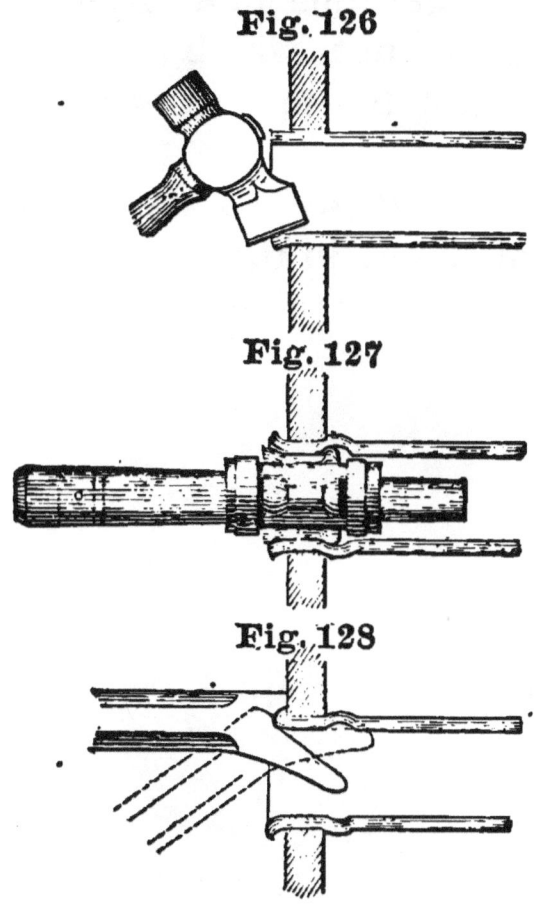

Fig. 126

Fig. 127

Fig. 128

spaces between them comparatively great, a number of them are fitted with nuts on both ends. Such tubes should be double the thickness of the others to allow sufficient strength after the threads are cut. The threading must be done in a lathe, and the sides of the nuts which bear upon the plate must be squared up with the thread.

Figures 126 to 128 show the three general stages in the progress of the work of setting tubes.

XVII.

FITTINGS.

Cast-iron, being so readily made to shape, is almost too frequently used for fittings. At the same time there is a vast deal of misconception of the proper proportions of the parts. A peculiar and inherent defect of cast-iron is the "shrinkage strain" produced by the unequal cooling of the metal in casting. To reduce this to a minimum, all portions of the same piece should be as nearly of the same weight as possible, or the difference be not abrupt. For example, pipes, as usually made, have the flanges from one and one half to twice the thickness of the barrel, and with but an ordinary fillet in the corner. If the flange be broken off, it will be found honeycombed and weak. This may be avoided almost entirely by making the fillet of a parabolic shape, as in figure 129. Or, if the room is restricted, it may be made of the same thickness as the barrel, and a number

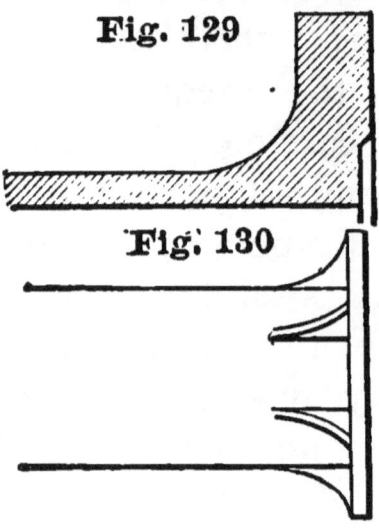

Fig. 129

Fig. 130

of short brackets cast in, as in figure 130.

Supporting lugs for cylindrical boilers are troubled the same way if there is much variation in the thickness of the metal. The placing of these lugs should be such as to produce as little local strain upon the boiler as possible. The best practice places them not over four diameters apart. Where two pairs only are required, place them one sixth of the length from each end. A riding plate of considerable area should always be placed on top of the brick-work for the lugs to rest upon; and all but one pair, be there many or few, should be provided with friction rollers. There should be a liner inside of the boiler beneath every supporting lug. It should be riveted at the corners, independently of the rivets that hold the lug.

Manholes should be kept as small as possible because of the cutting away of the boiler's proper strength. Oval 12″ × 15″ is a good size, and while very large men cannot enter them, those considerably above the average girth can do so by raising their arms above their heads. Where the bracing does not take up so much room as to prevent access, the ring should invariably be put inside both for strength and for convenience in calking. It is hardly necessary to say that the plate should be inside so that the pressure of the steam will help to keep the joint tight. Where a soft gasket (rubber, asbestos, and the like) is used, it is not necessary to have the joint machined if fairly straight and smooth. The plate should have two handles, whether there be two bolts or one. No one who has not wrestled with a single-handled plate can conceive the amount of exasperation caused by it. Both handles and bolts can and should be cast into the plate, the bolt having a roughened head being held in a generous boss.

Hand holes do not need a ring, but make their seating directly upon the plate. They should never be used where the seating is of necessity partly curved and partly flat, as it requires the greatest amount of patience and trouble to get them tight, and are a constant annoyance forever.

Mud plugs should be used in such places. These should be quite tapering, and, when new, should make the joint when the points have just shown through. The full length should be two and a half to three times the thickness of the plate to allow for re-tapping in case the thread of the hole is injured by the bars when cleaning.

Fusible or safety plugs should be placed just on the line where the fire reaches highest and can play directly upon them : in a tubular boiler, in the back uptake, central, and just above the tubes; in boilers of the locomotive type, in the fire-box crown, not far from the tube plate. It is desirable in such a situation that there be a cleaning hole opposite it so that it can be cleaned of scale or mud, which is apt to cover it there. Scale frequently (with bad water) forms so hard and thick, that although the tin may be melted out of the plug no water or steam can get through, in which case it might better be termed a "danger plug" instead of safety plug.

The best material for filling safety plugs is "Banca tin," unalloyed.

The hole in the plug should be tapering, with the large end outward so

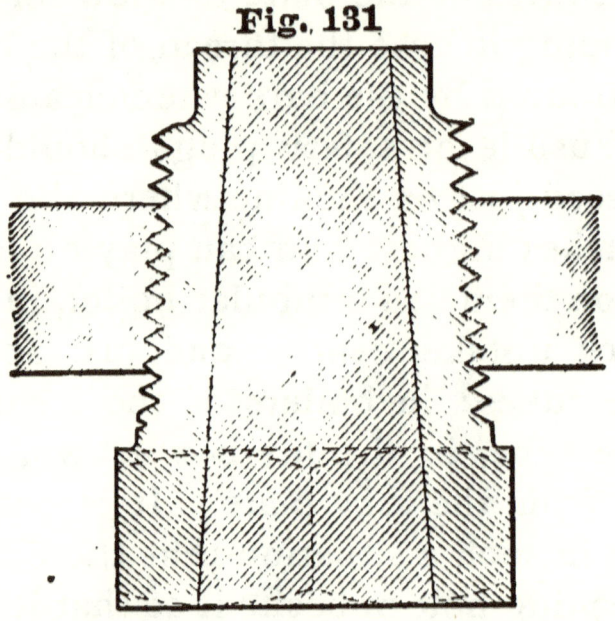

Fig. 131

as to present the least obstruction to the fused metal when it starts. Figure 131 is a full-sized section of a plug that has given good service wherever used.

Holes for gauge cocks should be

cut so that there will be not less than 2½″ height between the *highest point* reached by the fire and the *lowest* cock. 3″ is a better figure, because of defects in making and setting the boiler.

All holes tapped for removable fittings should be tapering to allow the joint to be made in the thread. The joint should never be made with a shoulder in such cases, for but very few removals will destroy the thread so that it will not draw up, whereas if it depends on the tapering thread it becomes compressed, and in consequence has a greater endurance.

Gaspipe taps are about the best thing for such work in the smaller sizes, but while the taper and diameter of the larger sizes are convenient, the pitch of the thread is too coarse. It should never be greater than eleven and a half per inch to insure good work. Less than three threads in the plate is very unreliable.

XVIII.
TESTING.

THERE are various methods of testing boilers before leaving the shop, and each has its adherents, though it would seem that the ideal test is that one which approaches closest to the conditions of actual service. Save for those with internal fire-boxes, this is of course impracticable; but for such it is the proper thing.

The distortion caused by the localization of the intense heat shows at once all the little leaks and weak spots, which would otherwise cause words of cerulean tint to float around the fire-door during the boiler's infancy. The only disadvantage of this method is the heat arising from the ashes and grate bars after the fire is drawn, which makes it almost impossible for the men to enter. However, this can be obviated by using a temporary knockdown grate, which may be hauled out with a poker or other hooked instrument. To catch the ashes a plate of light iron can be run under the fire-box, if the box be of the open bottom variety, which can be pulled out before the men enter. The same thing can be done with a close bottom if care is taken not to fit it too close. The warping caused by the heat will fasten it in otherwise.

The cold water, or hydrostatic test, is decidedly hurtful, except in the hands of the most careful. The vibrations caused by the sudden starting and stopping of the flow of water from the pump is only too apt to "rattle" the boiler to such an extent that what was in reasonably fair condition will be worse than the poorest.

Where sufficient draught cannot be had to produce a service test recourse may be had to a warm-water pressure. This is done by completely filling the boiler with water. This filling must be done from the highest point, so as to leave no contained air. Or if water is taken from a hydrant, leave the safety-valve open until the water is well out of it. Air gives no show when leaking through the joints as water will. Build but a light fire, and shortly the boiler will give indications of labor. Great care must be taken in working the fire in this test, as the expansion of the water is positive, and unlike steam is inelastic. It is both powerful and rapid when once it begins, being but slightly less than 5 cubic feet for every hundred contained by the boiler between the temperatures of 32° and 212°, or very nearly $\frac{1}{40}$ cubic foot for every degree of heat.

The advantage of this over the

cold-water pump test is that it is even and constant, consequently there is no shock to "rattle" the boiler.

Frequently, it will be noticed that quite a number of leaks will show, from the bare weight of the water about the lower part of the boiler. A few of such leaks will close up as the heat is raised, but it is much better to close them all as they show than to wait and find out after the boiler is heated. The work is disagreeable enough at the best, and a little time lost in closing such little cold-water affairs will more than repay itself.

Another method is to connect the boiler to a stationary one in which the steam is raised. Sometimes steam is taken from the shop-boiler, but this has a bad effect on the machinery, and besides, it requires an extra strength in a large boiler that is of no other use. A small auxiliary boiler, with capacious grate and large heating surface, to be used only for heating purposes, is an indispensability in all shops where good work is done.

Fill the boiler to be tested about one third full of water before blowing

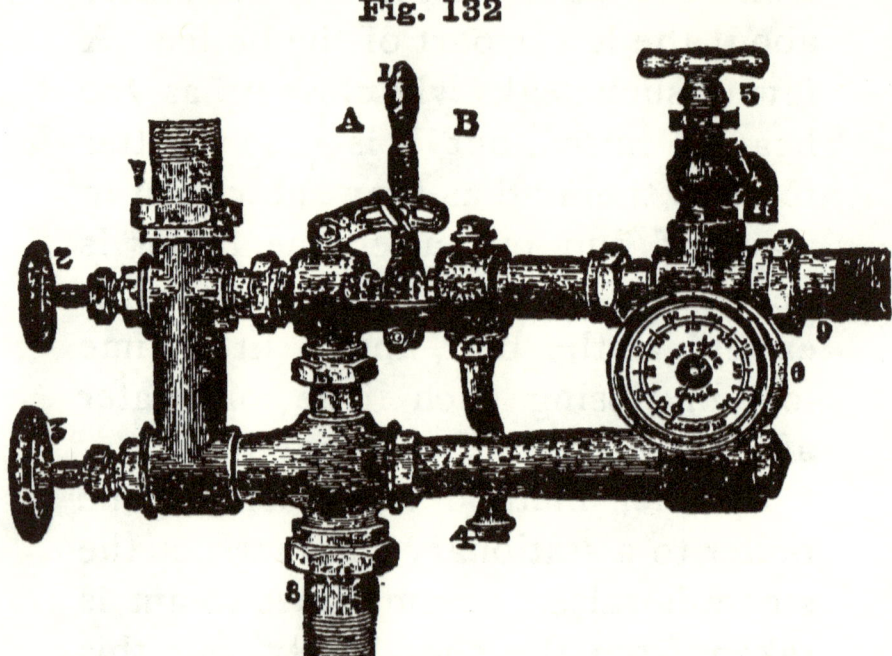

Fig. 132

in the steam. Much more or less will add considerably more to the quantity of steam required from the fired boiler, and steam means fuel.

Barring a service test, the best practical method now employed is by use of the "Little Giant" boiler-testing apparatus, shown in figure 132, which fills the boiler with hot water and then applies the pressure gradually, without shock, and holds it there.

The arrangement is a combination of two distinct instruments: the upper one in the cut is an ordinary injector which fills the boiler; the lower one is used for raising the pressure.

In setting up this apparatus, connect 7 to the steam and 8 to the water pipes, while 9 leads to the boiler to be tested. It must be kept in mind that as this is not a "lifting" injector, the water must flow freely to it.

To operate, move 1 to *B*, then turn on the water full. Next open valve 3 gradually, until wide open. When the boiler is full close it (3) again. To apply the pressure, move lever 1 to a perpendicular position, as shown in the cut. When the water flows from the overflow 4, open the valve

2 gradually until wide open. Then if steam or water shows at the overflow 4, regulate by moving the lever 1 toward *A* or *B*, as may be required. The pressure obtained will be indicated by the gauge. To fix the pressure, screw handle 5 up or down—down to increase, up to release.

These instruments are made regularly to produce any pressure desired up to three times the initial steam pressure used, and if required can be made to go much higher.

There should, invariably, be a gauge on the boiler under test, and this gauge should be so placed as to be entirely out of the way of the inflowing steam or water—whichever is used. It is highly important that such gauges be accurate and reliable. Frequent test should be made of them. Straining from the high pressures put upon them is apt to weaken them, and in consequence there is not nearly so much pressure on the boiler as they indicate. Should they happen to indicate, say, 100 lbs. when there was but 70 lbs., and the boiler is fixed up right and tight to all appearances, when in fact it is not, when put in service there is a woful time for maker or user, or mayhap both.

XIX.

ORDERING STOCK.

The first thing to be done on receipt of the drawing is to make up a bill of materials requisite. This includes, besides the plating, tubes (if called for), rivets, stay bolts and rods, plate and bars for stays, bar iron for water-legs, etc.

The plate iron should be gotten out, as it requires some time to get it from the mill when it is rolled to size. There are standard sizes of plates kept in stock, but it is rarely that they will cut to the size wanted without considerable waste. It would be well to lay the boiler full size, if of irregular shape, thereby checking the drawing and reducing the chances of keeping

odd plates in stock. In getting up
this bill, keep such plates as are of the
same material and thickness together.
Give also each item a specific mark,
either letter or number, irrespective
| 12 | 1 | Back Head | Ex. Fl. | $87\tfrac{1}{2}/_{50} \times 55/_{49}$ | $\tfrac{1}{2}$ | Place for sketch. |
of the number of plates to each. This
will prove very convenient in case of
dispute, as the particular item can be
identified at once.

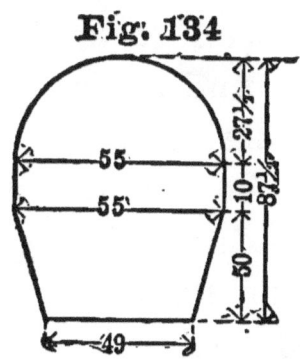

Fig. 134

Plates of other than rectangular or
circular shape must have a little figured sketch accompanying. In the
dimensions given for irregular shapes
generally the two principal ones each
way are given. Figure 134 is the
sketch of the back head of a locomo-

tive boiler. The item accompanying it is given in full.

Always give that dimension first that runs the direction in which the plate is to be bent. This is absolutely necessary only in lower grades than extra flange. With extra flange and higher grade it is not so requisite, because such are supposed to be capable of bending in any direction. Yet, if it is not necessary in high grade stock, it is well to make a regular thing of it, and there will be less chance of missing it when ordering low grade. In cutting out plates in the mill, the dimension given first is laid off in the direction in which the plate passed the rolls, and as in low grade stock that is the direction in which the fibers form, its importance is self-evident. As an example, take the welt plates for a butt-jointed seam; if it is written, say, 4 plates—shell—$50 \times 10\tfrac{1}{2} \times \tfrac{5}{8}$, you will surely have stock that will inevitably split on bending; but if written, 4 plates—shell—$10\frac{1}{2} \times 50 \times \frac{5}{8}$, there is little fear of such consequences.

Long rectangular plates may be ordered very closely to the finished width, but an inch or two should be allowed in the length. The squares used in rolling mills are very apt to be something else. Allow plenty of trimming on irregularly shaped plates, as it is much better to lose a little in waste than to risk a great deal.

Rivets being sold by the pound, irrespective of their size, it necessitates, after determining the number required, calculating their weight. All manufacturers furnish tables of weight of 100 rivets; therefore, to find the weight required, divide the number of rivets by 100 and multiply by the weight given for that particular size and length. For example, seven hundred and fifty-four $\frac{3}{4}''$ rivets $1\frac{7}{8}''$ long will weigh $754 \div 100 \times 36.70$ (see table at end of this book) $= 276.7$ pounds.

It is best to add about ten per cent to allow for waste by burning, bad driving, etc. Call it in this case 300 pounds.

Round and flat iron can be ordered by the running foot.

WEIGHT OF IRON AND STEEL PLATE.

Thickness in Inches.	Iron.		Steel.	
	Pounds per sq. foot.	Pounds per sq. inch.	Pounds per sq. foot.	Pounds per sq. inch.
1/8	5.05	.0351	5.08	.0353
3/16	7.58	.0526	7.62	.0529
1/4	10.10	.0701	10.16	.0707
5/16	12.63	.0877	12.70	.0882
3/8	15.16	.1053	15.24	.1057
7/16	17.68	.1228	17.78	.1235
1/2	20.21	.1403	20.32	.1411
9/16	22.73	.1578	22.86	.1588
5/8	25.26	.1754	25.40	.1764
11/16	27.79	.193	27.94	.1926
3/4	30.31	.2105	30.48	.2117
13/16	32.84	.228	33.02	.2193
7/8	35.37	.2456	35.56	.1636
15/16	37.89	.2631	38.10	.2646
1	40.42	.2807	40.65	.2823

CONE-HEAD BOILER RIVETS.
Weight per Hundred.
HOOPES & TOWNSEND.

Length.	½ inch.	⅝ inch.	11⁄16 inch.	¾ inch.
¾ inch.	8.75	16.20	.....	.....
⅞ "	9.35	17.22	.....	.....
1 "	10.00	18.25	21.70	26.55
1⅛ "	10.70	19.28	23.10	28.00
1¼ "	11.40	20.31	24.50	29.45
1⅜ "	12.10	21.34	25.90	30.90
1½ "	12.80	22.37	27.30	32.35
1⅝ "	13.50	23.40	28.70	33.80
1¾ "	14.20	24.43	30.10	35.25
1⅞ "	14.90	25.46	31.50	36.70
2 "	15.60	26.49	32.90	38.15
2⅛ "	16.30	27.52	34.30	39.60
2¼ "	17.00	28.55	35.70	41.05
2⅜ "	17.70	29.58	37.10	42.50
2½ "	18.40	30.61	38.50	43.95
2⅝ "	19.10	31.64	39.90	45.40
2¾ "	19.80	32.67	41.30	46.85
2⅞ "	20.50	33.70	42.70	48.30
3 "	21.20	34.73	44.10	49.75
3¼ "	22.60	36.79	46.90	52.65
3½ "	24.00	38.85	49.70	55.55
3¾ "	25.40	40.91	52.50	58.45
4 "	26.80	42.97	55.30	61.35
4¼ "	28.20	45.03	58.10	64.25
4½ "	29.60	47.09	60.90	67.15
4¾ "	31.00	49.15	63.70	70.05
5 "	32.40	51.21	66.50	72.95
5¼ "	33.80	53.27	69.20	75.85
5½ "	35.20	55.33	72.00	78.75
5¾ "	36.60	57.39	74.80	81.65
6 "	38.00	59.45	77.60	84.55
6½ "	40.80	63.57	83.30	90.35
7 "	43.60	67.69	88.90	96.15

TANK RIVETS.
Number to the Pound.
HOOPES & TOWNSEND.

Length.	3/16 in. Diam.		1/4 in. Diam.		5/16 in. Diam.		3/8 in. Diam.	
	F. & R. H.	C. S.	F. & R. H.	C. S.	F. & R. H.	C. S.	F. & R. H.	C. S.
3/8	165	230	103	155	67	89	..	..
7/16	153	200	92	128	58	78	35	..
1/2	135	172	81	108	50	69	31	46
5/8	118	148	71	93	44	60	27	40
3/4	103	129	63	80	39	53	23	35
7/8	92	114	56	70	35	47	21	31
1	84	102	50	62	32	42	19	27
1 1/8	77	93	46	56	30	37	18	24
1 1/4	72	85	43	51	28	34	17	22
1 3/8	67	78	40	47	26	31	16	21
1 1/2	62	72	37	44	24	29	15	20
1 5/8	58	67	34	41	22	27	14	19
1 3/4	54	62	32	38	21	25	13	18
1 7/8	51	58	30	35	20	23	12	17
2	48	54	29	33	19	22	11	16

F. & R. H. indicates Flat and Round Heads. C. S. indicates Countersunk.

INDEX.

[In looking for any subject having more than one word, look first for what appears to you to be the principal word, and if you do not find it under the initial letter of that word, try under that of the other.]

	PAGE
Angle of calking edge	155
Angle iron for brace feet	80, 82
ANNEALING	115
Annealing furnace	117, 118, 119
Annealing without a furnace	116
Anvil for forging crow-foot braces	134
Back connection marine boiler	30
Bad joint	47
Bar for holding on	53
Bar iron, ordering	208
Bars for crown plate of furnaces	70
Beading tool	186
Beading tubes, length to allow for	182
"Belpair" furnace	37
BENDING, COLD	159
Bending plate with large openings	162
Bending test for plate	17
Bent brace jaws	84
Bill of materials	204

	PAGE
Blisters...............................	43
Block for holding on in narrow space......	54
BOILER FORMS...........................	20
Boiler, turning over.....................	172
Bolts, clamping.........................	168
Bolts, crown sheet......................	68
Bolts, manhole plate....................	192
Brace feet..............................	80
Brace feet, flanged angle iron for..........	82
Brace jaws.........................77,	128
Brace jaws, crooked.....................	84
Brace jaws, forging.....................	128
Braces and stays, direction of.............	59
Braces and stays, fitting..................	59
Braces and stays, number and size.......58,	85
Braces, crow-foot....................80,	134
Braces, for large boilers..................	85
Braces, pins for.........................	78
BRACING AND STAYING....................	58
Brands of iron plate...................10,	12
Bursting strength of cylinders.............	22
Butt joints.........................26,	45
Button rivet point.......................	51
CALKING	175
Calking, concave...................176,	178
Calking, evil effects of over...............	175
Calking, styles of........................	175
Calking tools.......................176,	180

INDEX. 215

	PAGE
Cam lever for bending crown bars	125
Castings for formers	93
Cast iron	14
Cast iron fittings	189
Cast iron heads	14
Cast iron pipe	189
Cast iron pipe flanges	189
Cast iron shrinkage strains	189
Chain, jack, for setting up	166
C. H. No. 1, Fire box	10, 11
C. H. No. 1, Flange	10, 11
Circular heads	33
Circular heads, trimming	157
Clamping bolts	168
Clamping plates for flanging	98
Clamping sides of fire boxes	170
Cocks, fitting holes for gauge	194
Cocks, position of gauge	195
COLD BENDING	159
Cold bending plate with large openings	162
Cold bending rolls	160
Cold, effect of, on rivets	57
Cold water, or hydrostatic test	197
Combustion chambers	41, 42
Compensating rings	31
Concave calking	176, 178
Conical rivet point	51
Copper ferrules	182
Copper fire boxes	9

	PAGE
Corners, smithing of plate	122
Corners, trimming of plate	157
Corrosion of stays	61
Counterboring tool	152
Countersinking tool	51, 151
Countersunk rivets	51
Crane, traveling	164
Crooked brace jaws	84
Crow-foot brace	80, 134
Crow-foot brace, anvil for forging	134
Crow-foot brace, forging	134
Crown bar	70
Crown bar bolts	68
Crown bar, cam lever for bending	125
Crown bar, fitting	71
Crown bar, former	124
Crown bar, formula	73
Crown bar heating furnace	126
Crown bar slings	74
Crown bar smithing	123
Crown bar thimble	71
Crown sheet bolts	68
Curved crown bars	123
Cylinder, annealing	117
Cylinder bursting strength	22
Cylinder, external pressure on	25
Cylinder, force to burst transversely	24
Cylinder thickness	21
Cylinder welding	111

INDEX.

	PAGE
Cylindrical boiler, setting up	174
Cylindrical flues	27
Cylindrical form of boiler	20
Die formula	147
Dies for punches	147
Direct fire the proper test	196
Direction of stays and braces	59
Disposition of stays and braces	58
Distortion from heat	196
Dome flanging	107
Dome opening, bending plate with	162
Dome, welted opening of	32
Domes	31
Domes, setting up	172
Doors, furnace	40
Drawing of the metal in flanging	105
Dressing formers	95
Drifting rivet holes to bring fair	54
Drilled rivet holes	56
DRILLING, ETC.	150
Drilling machine	153
Drills, twist	150
Driving rivets	52, 55
Ductility of metal	10, 13
Effect of cold on riveting	57
Effect of overdriving on calking	175
Elongation of iron in test	12

	PAGE
Elongation of steel in test	10
English method of welding plates	109
Expanding tool	184
Expanding tubes	183
External pressure on cylinders	25
Faces of punches	144
Factor of safety	21
Falls for hoisting and lowering	173
Feet of braces	80
Fitting braces and stays	59
Fitting crown bars	71
FITTINGS	189
Fittings, cast iron	189
Fitting screwed stays	62
Fire boxes, clamping in sides	170
Fire boxes, copper	9
Fire boxes, setting up	169
Fire box, plumbing and lining	172
Fire box rings, forging	135
Fire box rings, slotting	137
Fire door flange	93
Fire, flanging	95
Fire welding	113
Flanged plates, sling for	96
Flanged tee for brace feet	81
Flanged tee former	126
Flanged tee-iron	127
Flanged work, trimming	158

INDEX.

	PAGE
Flange fire	59
Flange fire tuyere	95
Flange former pattern	93
Flange formers	92
Flange heating	96
Flange iron	16
Flange iron testing	17
Flange, C. H. No. 1	10
Flange, radius of	92
Flanges on cast iron pipes	189
FLANGING	92
Flanging, clamping plate while	98
Flanging domes	107
Flanging forge, height of	96
Flanging gauge	104
Flanging, holding down staves for	103
Flanging mauls	100
Flanging methods	92
Flanging over anvil or block	103
Flanging under black heat	102
Flat ends	24
Flat surfaces	33
Flatter handles	101
Flatters	101
Flue joints, furnace	28, 35
Flues, thickness of cylindrical	27
Flues, transverse joints	27, 28
Forces to burst cylinder transversely	24
Forging brace jaws	128

INDEX.

	PAGE
Forging crow-foot brace	134
Forging curved crown bars	123
Forging fire box rings	135
Forging flanged tee	127
Former castings	93
Former for flanging	92
Former for smithing crown bars	124
Forming brace jaws	132
FORMS OF BOILERS	20
Formula, circular heads	33
Formula, crown bars	73
Formula, cylinder to resist bursting	22
Formula, force to burst cylinder	24
Formula, longitudinal strength of cylinder	24
Formula, punch dies	147
Formula, riveted joints	48
Formula, stays	61
Formula, thickness of cylinder	21
Formula, thickness of cylindrical flues	27
Furnace, annealing	117
Furnace, "Belpair"	37
Furnace doors	40
Furnace for heating crown bars	126
Furnace flue joints	28, 35
Furnace plates	42
Furnaces	34
Furnaces, locomotive	35
Fusible plugs	193
Fusible plugs, material for filling	194

	PAGE
Gasket for manholes	192
Gas-pipe taps	195
Gauge cocks	194
Gauge cocks, position of	195
Gauge cocks, taps for	195
Gauge, flanging	104
Gauge, testing	202
Girth seam riveting	52
Grate, temporary for testing	197
Grooving	46
Guarantee of strength	13
Gusset stays	88
Hammer, holding-on	53
Hand holes	192
Handles, manhole plate	192
Hand riveting	53
Heads	14, 34
Heads, cast iron	14
Heads, circular	33
Heads, trimming circular	157
Heating crown bars	126
Heating flanges	96
Heating rivets	53
Heating steel	17, 114
Heavy braces	85
Heavy braces, turnbuckle for	86
Heavy plate punching	143
Height of flange forge	96

INDEX.

	PAGE
Hoisting and lowering boilers	172
Holding-down staves for flanging	103
Holding-on bar	53
Holding-on block for narrow space	54
Holding-on hammer	53
Holes for tubes, punching	148
Hollow screwed stays	65
Hooks, porter, for handling plate	99
Hornbeam	100
Hydrostatic test	197
Injector for testing	200
Inspectors' rate for iron	12
Iron, cast	14
Iron, elongation in test	12
Iron for rivets	13
Iron for stay bolts	13
Iron plate	10
Iron plate, brands of	10
Iron plate, weight of	209
Iron, refined	11
Iron, rolled tee	29
Iron, shell	11
Iron, stay rod	14
Iron, tank	11, 12
Iron, tensile strength	11, 14
Iron, testing flange	17
Iron, U. S. Gov. inspectors' rate	12
Iron, welding heat for	114

INDEX.

	PAGE
Jack chain for setting up	166
Jaws of braces	77, 84, 127
Jaws of braces, bent	84
Joint, bad	46
Joint, butted and welted	26, 45
Joints, formulas for riveted	48, 49
Joints, furnace flue	27, 28, 35
Joints, lap	26, 44
Joints, longitudinal	26
JOINTS, RIVETED	44, 47
Joints, table of proportions of riveted	49, 50
Joints, transverse of flues	27, 28
Joints, welted	45
Lamination of plates	15
Lap joints	26, 43
Lap welded tubes	30
Lap welding plates	110
Large braces	85
Large opening, cold bending plate with	162
Large tubes, setting	187
Leaks, small	199
Length to allow for beading	182
Lining up	167
Lining up with straight edge	168
"Little Giant" boiler testing apparatus	200
Locomotive furnaces	36
Locomotive wagon top	36
Longitudinal lap joint	26

	PAGE
Longitudinal strength of cylinders	24
Long screwed stay	65
Long screwed stay, support for	67
Long screwed stay, tap for	66
Lowering with a fall	173
Lugs for crown bar slings	76
Lugs, riding plate for supporting	191
Lugs supporting, for cylindrical boilers	190
Machines for driving rivets	52
Machines for trimming plate	156
Manholes	31, 162, 191
Manholes, gaskets for	192
Manhole opening, bending plate at	162
Manhole plate bolts	192
Manhole plate handles	192
Manholes, size of	191
Marine back connection	31
Marine boiler	34
MATERIALS	9
Materials, bill of	204
Material for filling safety plugs	194
MATERIALS, TESTING	14
Marking-off for punching	140
Mauls for flanging	100
Measure for length of tubes	181
Methods of flanging	92
Methods of testing	196
Mud plugs	193

INDEX.

	PAGE
Narrow water spaces	37
Narrow water space stays	60
N. P. U. iron	12
Number and size of braces and stays	58
Openings for domes	31, 33
Ordering bar iron	204, 208
Ordering plates	204
Ordering rivets	207
ORDERING STOCK	204
Overdriving in calking	175
Pattern, flanging former	93
Pine iron	12
Pins, brace	78
Pipe flanges, cast iron	189
Pipes, cast iron	189
Plate, brands of iron	10
Plate bending test	17
Plate edge trimming machine	156
Plate, iron	10
Plate, ordering	204
Plate, riding, for supporting lugs	191
Plate, punching heavy	143
Plate, smithing corners	121
Plate, steel	10
Plate, testing	15
Plate test, table of angles bending	16
Plate, weight of iron and steel	209
Plate welding	108

	PAGE
Plate, working test	16
Plates furnace	42
Plates, lap welding	110
Plumbing fire box	172
Plugs, fusible or safety	193
Plugs, material for filling safety	194
Plugs, mud	193
Pockets in fire box sides	137
Points of rivets	50
Porter hooks	99
Position of gauge cocks	195
Power riveting	52
Pressure, external on cylinder	25
Proportions of riveted joints, table	49
Punch die, formula	147
Punch die	147
Punch faces	144
Punched hole, properly fitted	149
PUNCHING	138
Punching heavy plate	143
Punching holes for tubes	148
Punching, marking off for	140
Punching table	139
Punch, spiral shearing	144
Punch stocks	146
Radius of flanges	92
Record of strength	13
Rectangular boilers	34

INDEX.

	PAGE
Refined iron	11
Riding plate for supporting lugs	191
Rimers	54, 154
Rivet driving	55
RIVETED JOINTS	44
Riveted joints, designing	47
Riveted joints, formulas	48, 49
Riveted joints, table of proportions	49, 50
Rivet holes, drilled	56
Rivet holes, drifting to bring fair	54
RIVETING	50
Riveting, effect of cold on	57
Riveting girth seams	52
Riveting, hand	53
Riveting, power	52
Riveting with a snap	56
Rivet iron	13
Rivet iron strains	13
Rivet points	50
Rivets, heating power driven	53
Rivets, heating hand driven	54
Rivets, ordering	207
Rivets, steel	13
Rivets, weight of, table	210, 211
Rivet, test	19
Rolled tee iron	29, 81
Rolling plate, cold	161
Rolls for bending cold	160
Rounding tool, tube hole	153

INDEX.

	PAGE
Safety, factor of	21
Safety plugs	193
Safety plugs material for filling	194
Screwed stays	60, 65
Screwed stays, fitting	62
Screwed stays, formula	61
Screwed stays, hollow	65
Screwed stays, support for long	67
Screwed stays, taps for	63, 66
Screwed stays, wrench for	64
Setting copper ferrules	183
Setting large tubes	187
Setting screwed stays, wrench for	64
SETTING TUBES	181
SETTING UP	164
Setting up cylindrical boilers	165, 174
Setting up domes	172
Setting up fire boxes	169
Setting up floor	164
Setting up locomotive boilers	169
Shell iron	11
Shrinkage strains in cast iron	189
Single riveted lap-joint	44
Size and number of braces and stays	58
Size of crown bar	73
Size of manholes	191
Sligo iron	12
Slings, crown bar	74
Slings for carrying plates	96

Slotting fire box rings	137
Small leaks	199
SMITHING	121
Smithing crown bars	123
Smithing fire box rings	135
Smithing plate corners	121
Snap, riveting with	56
Special brands of plate	12
Spherical form of boiler	20
Spiral shearing punch	144
Stay bolt iron, tensile strength	13
Stay bolt iron, test	19
Stay bolts	60
Stay bolts, crown sheet	68
STAYING AND BRACING	58
Stay rod iron	14
Stays and braces, direction of	59
Stays and braces, fitting	59, 62
Stays and braces, number and size	58
Stays, corrosion of	61
Stays, fitting screwed	62
Stays, formula	61
Stays, gusset	88
Stays, hollow screwed	65
Stays, long screwed	65
Stays, taps for screwed	62, 66
Stays, throat plate	87
Stays through plate at angle	67
Stays, support for long	67

INDEX.

	PAGE
Stays, wrench for screwed	64
Staves for holding down while flanging	103
Steady pins for flanging	95
Steel, elongation of in test	10
Steel, heating	17, 114
Steel plate	10
Steel plate, weight of, table	209
Steel rivets	13
Steel rivets, tensile strength of	13
Steel, strength of	10
Steel, testing	17
Steel, test piece	10
Steel, welding heat	114
Stiffening flat surfaces without bracing	83
Stock ordering	204
Stocks, punch	146
Stop-water	27, 175
Straight edge, lining up with	168
Strains on rivet iron	13
Strength of clyinder	22, 24
Strength of steel	10
Styles of calking	175
Supporting lugs	190
Supporting lugs, riding plate for	191
Support, long screwed stay	67
Surfaces, flat stiffened without bracing	83
Swedging tubes	183
Table, angles for bending plate in test	16
Table for front of punching machine	140

INDEX. 231

	PAGE
Table, proportions of riveted joints	49, 50
Table, weight of cone head rivets	210
Table, weight of iron and steel plate	209
Table, weight of tank rivets	211
Taking measure for length of tubes	181
Tank iron	11, 12
Tank rivets, weight of	211
Taps for screwed stays	62, 66
Taps, gas pipe	195
Taps, gauge cock	195
Tee-iron for brace feet	81
Tee iron flanged	81, 126
Tee iron rolled	29, 81
Templet, for marking off for punching	142
Temporary grate for testing	197
Tensile strength	9, 12
Tensile strength of iron	11
Tensile strength of stay-bolt iron	13
Tensile strength of stay-rod iron	14
Tensile strength of steel	10
Tensile strength of steel rivets	14
Test gauge	202
Test by steam from another boiler	199
Test by steam, injector forced	200
Test, hydrostatic or cold water	197
Test, warm water	198
Test for plate, bending	17
TESTING	196
Testing, filling boiler for	198, 200
Testing flange iron	17

	PAGE
TESTING MATERIALS	14
Testing plate	15
Testing, temporary grate for	197
Test piece, steel	10
Test rivet	19
Test, stay-bolt iron	13, 19
Test, steel	17
Test, tube	18
Thickness of cylinder	21
Thickness of cylindrical flues	27
Thickness of plate, testing	15
Thimble, crown bar	71
Throat plate	105
Throat plate stays	87
Tool for beading tubes	186
Tool for counterboring	152
Tool for countersinking	51, 151
Tool for expanding tubes	184
Tool for holding tubes in place	184
Tool for rounding tube holes	153
Toughness of metal	10, 13
Transverse joints of flues	27
Traveling crane	164
TRIMMING	155
Trimming circular heads	157
Trimming corners	157
Trimming flanged work	158
Tube holes, punching	148
Tubes	9, 18, 30, 181

INDEX.

	PAGE
Tubes, expanding	183
TUBE, SETTING	181
Tubes, setting large	187
Tubes, small lap-welded	30
Tubes, swedging	183
Tubes, taking measure for length	181
Tubes, tool for beading	186
Tube test	18
Turnbuckle for heavy braces	86
Turning over boilers	172
Tuyere, flanging fire	95
Twist drill	150
U. S. Gov. inspectors' rate for iron	12
Wagon top	36
Warm water test	198
Water leg bottoms	38
Water spaces, narrow	37
Water space stays, narrow	60
Weight of iron and steel plate	209
Weight of rivets	210, 211
Welding cylinders	111
Welding fire	113
Welding heat	114
Welding jaws to braces	128
WELDING PLATES	108
Welding plates, English method	109
Welted joints	45
Wrench for screwed stays	64

www.ingramcontent.com/pod-product-compliance
Lightning Source LLC
Chambersburg PA
CBHW021811230426
43669CB00008B/707